Springer Proceedings in Physics

Volume 272

Indexed by Scopus

The series Springer Proceedings in Physics, founded in 1984, is devoted to timely reports of state-of-the-art developments in physics and related sciences. Typically based on material presented at conferences, workshops and similar scientific meetings, volumes published in this series will constitute a comprehensive up-to-date source of reference on a field or subfield of relevance in contemporary physics. Proposals must include the following:

- name, place and date of the scientific meeting
- a link to the committees (local organization, international advisors etc.)
- scientific description of the meeting
- list of invited/plenary speakers
- an estimate of the planned proceedings book parameters (number of pages/articles, requested number of bulk copies, submission deadline).

Please contact:

For Americas and Europe: Dr. Zachary Evenson; zachary.evenson@springer.com
For Asia, Australia and New Zealand: Dr. Loyola DSilva; loyola.dsilva@springer.com

More information about this series at https://link.springer.com/bookseries/361

Kang-Ping Lin · Ren-Shyan Liu · Bang-Hung Yang
Editors

Future Trends and Challenges of Molecular Imaging and AI Innovation

Proceedings of FASMI 2020

Springer

Editors
Kang-Ping Lin
Department of Electrical Engineering
Chung Yuan Christian University (CYCU)
Taoyuan, Taiwan

Bang-Hung Yang
Department of Biomedical Imaging
and Radiological Sciences (BIRS)
National Yang Ming Chiao Tung University
(NYCU)
Taipei, Taiwan

Ren-Shyan Liu
Department of Nuclear Medicine
and Department of Biomedical Imaging
and Radiological Sciences (BIRS)
Cheng Hsin General Hospital and National
Yang Ming Chiao Tung University (NYCU)
Taipei, Taiwan

ISSN 0930-8989 ISSN 1867-4941 (electronic)
Springer Proceedings in Physics
ISBN 978-3-030-92785-1 ISBN 978-3-030-92786-8 (eBook)
https://doi.org/10.1007/978-3-030-92786-8

This Springer imprint is published by the registered company Springer Nature Switzerland AG
The registered company address is: Gewerbestrasse 11, 6330 Cham, Switzerland

Contents

Contributors

Najim Akhtar Institute of Biophotonics, National Yang Ming Chiao Tung University, Taipei, Taiwan

Hui-Wen Chan Department of Biomedical Imaging and Radiological Sciences, National Yang Ming Chiao Tung University, Taipei, Taiwan

Chi-Wei Chang Department of Nuclear Medicine and National PET/Cyclotron Center, Taipei Veterans General Hospital, Taipei, Taiwan

Chih-Hsien Chang Isotope Application Division, Institute of Nuclear Energy Research, Taoyuan, Taiwan

Deng-Yuan Chang Department of Biomedical Imaging and Radiological Sciences, National Yang Ming Chiao Tung University, Taipei, Taiwan

Teng-Bin Chang Department of Electrical Engineering, Chung Yuan Christian University, Taoyuan, Taiwan

Surojit Chattopadhyay Institute of Biophotonics, National Yang Ming Chiao Tung University, Taipei, Taiwan

Chao-Cheng Chen Department of Biomedical Imaging and Radiological Sciences, National Yang Ming Chiao Tung University, Taipei, Taiwan

Chuan Lin Chen Department of Biomedical Imaging and Radiological Sciences, National Yang Ming Chiao Tung University, Taipei, Taiwan

Ju-Yu Chen Department of Health Technology and Informatics, The Hong Kong Polytechnic University, Hong Kong, China

Mei-Fen Chen Department of Electrical Engineering, Chung Yuan Christian University, Taoyuan, Taiwan;
Technology Translation Center for Medical Device, Chung Yuan Christian University, Taoyuan, Taiwan

Ming-Wei Chen Isotope Application Division, Institute of Nuclear Energy Research, Taoyuan, Taiwan

Su-Jung Chen Isotope Application Division, Institute of Nuclear Energy Research, Taoyuan, Taiwan

Yi-An Chen Institute of Clinical Medicine, National Yang Ming Chiao Tung University, Taipei, Taiwan;
Molecular and Genetic Imaging Core, Mouse Clinic, National Comprehensive Mouse Phenotyping and Drug Testing Center, Taipei, Taiwan;
Department of Biomedical Imaging and Radiological Sciences, National Yang Ming Chiao Tung University, Taipei, Taiwan

Yi-Lun Chen Taipei Veterans General Hospital, National PET/Cyclotron Center and Department of Nuclear Medicine, Taipei, Taiwan;
Taipei Association of Radiological Technologists, Taipei, Taiwan

Chen-Han Cheng Institute of Nuclear Energy Research/Health Physics Division, Taoyuan, Taiwan

Hung-Wen Chiu Graduate Institute of Biomedical Informatics, Taipei Medical University, Taipei, Taiwan

Juri G. Gelovani Office of the Provost, United Arab Emirates University, Al Ain, United Arab Emirates

Sandip Ghosh Institute of Biophotonics, National Yang Ming Chiao Tung University, Taipei, Taiwan

Meilan Jimmy Hasugian Department of Electrical Engineering, Chung Yuan Christian University, Taoyuan, Taiwan

Wen-Sheng Huang Taipei Veterans General Hospital, National PET/Cyclotron Center and Department of Nuclear Medicine, Taipei, Taiwan

Yuan-Ruei Huang Isotope Application Division, Institute of Nuclear Energy Research, Taoyuan, Taiwan

Chien-Chih Ke Molecular and Genetic Imaging Core, Mouse Clinic, National Comprehensive Mouse Phenotyping and Drug Testing Center, Taipei, Taiwan;
Department of Medical Imaging and Radiological Sciences, Kaohsiung Medical University, Kaohsiung, Taiwan;
Department of Medical Research, Kaohsiung Medical University Hospital, Kaohsiung, Taiwan

Wing Sum Lau Department of Health Technology and Informatics, The Hong Kong Polytechnic University, Hong Kong, China

Shih-Ying Lee Isotope Application Division, Institute of Nuclear Energy Research, Taoyuan, Taiwan

Ming-Hsin Li Isotope Application Division, Institute of Nuclear Energy Research, Taoyuan, Taiwan

Hsin-Chin Liang Institute of Nuclear Energy Research/Health Physics Division, Taoyuan, Taiwan

Kang-Ping Lin Department of Electrical Engineering, Chung Yuan Christian University, Taoyuan, Taiwan;
Technology Translation Center for Medical Device, Chung Yuan Christian University, Taoyuan, Taiwan

Liang-Ting Lin Department of Health Technology and Informatics, The Hong Kong Polytechnic University, Hong Kong, China

Ren-Shyan Liu Program of Biomedical Science and Engineering, National Yang Ming Chiao Tung University, Taipei, Taiwan;
Institute of Clinical Medicine, National Yang Ming Chiao Tung University, Taipei, Taiwan;
Department of Biomedical Imaging and Radiological Sciences, National Yang Ming Chiao Tung University, Taipei, Taiwan;
Molecular and Genetic Imaging Core, Mouse Clinic, National Comprehensive Mouse Phenotyping and Drug Testing Center, Taipei, Taiwan;
Department of Nuclear Medicine, Cheng Hsin Hospital, Taipei, Taiwan;
Department of Nuclear Medicine and National PET/Cyclotron Center, Taipei Veterans General Hospital, Taipei, Taiwan

Sheng-Nan Lo Isotope Application Division, Institute of Nuclear Energy Research, Taoyuan, Taiwan

Shih-Wei Lo Isotope Application Division, Institute of Nuclear Energy Research, Taoyuan, Taiwan

Wei-Lin Lo Isotope Application Division, Institute of Nuclear Energy Research, Taoyuan, Taiwan

Cheng-Hsiu Lu Program of Biomedical Science and Engineering, National Yang Ming Chiao Tung University, Taipei, Taiwan;
Core Laboratory for Phenomics and Diagnostics, Kaohsiung Chang Gung Memorial Hospital, Kaohsiung, Taiwan;
Department of Medical Research, Kaohsiung Chang Gung Memorial Hospital, Kaohsiung, Taiwan

I-Ling Shih GE Healthcare, Taipei, Taiwan

Mao-Yuan Su Taipei Association of Radiological Technologists, Taipei, Taiwan;
Department of Medical Imaging, National Taiwan University Hospital, Taipei, Taiwan;
Department of Medical Imaging and Radiological Technology, Yuanpei University of Medical, Technology, Hsinchu, Taiwan

Wei-Fang Su Department of Materials Science and Engineering, National Taiwan University, Taipei, Taiwan;
Molecular Imaging Center, National Taiwan University, Taipei, Taiwan

Cheng-Chang Tsai Department of Materials Science and Engineering, National Taiwan University, Taipei, Taiwan

Cheng-Si Tsao Institute of Nuclear Energy Research, Longtan, Taoyuan, Taiwan

Yun-Hsiu Tseng Department of Materials Science and Engineering, National Taiwan University, Taipei, Taiwan

Hsin-Ell Wang Department of Biomedical Imaging and Radiological Sciences, National Yang Ming Chiao Tung University, Taipei, Taiwan

Shih-Min Wang Isotope Application Division, Institute of Nuclear Energy Research, Taoyuan, Taiwan

Hsiang-Ning Wu Institute of Nuclear Energy Research/Health Physics Division, Taoyuan, Taiwan

Ming-Chung Wu Department of Chemical and Materials Engineering, Chang-Gung University, Taoyuan, Taiwan

Ping-Yu Wu Department of Electrical Engineering, Chung Yuan Christian University, Taoyuan, Taiwan

Bang-Hung Yang Department of Nuclear Medicine and National PET/Cyclotron Center, Taipei Veterans General Hospital, Taipei, Taiwan;
Taipei Association of Radiological Technologists, Taipei, Taiwan;
Department of Biomedical Imaging and Radiological Sciences, National Yang Ming Chiao Tung University, Taipei, Taiwan

De-Ming Yang Microscopy Service Laboratory, Basic Research Division, Department of Medical Research, Taipei Veterans General Hospital, Taipei, Taiwan

Shun-Chung Yang Graduate Institute of Biomedical Informatics, Taipei Medical University, Taipei, Taiwan;
Taipei Association of Radiological Technologists, Taipei, Taiwan;
Department of Medical Imaging, National Taiwan University Hospital, Taipei, Taiwan

Tzu-Yi Yu Department of Materials Science and Engineering, National Taiwan University, Taipei, Taiwan

X-ray Tomography Study of 3D Hydrogel Structure

Cheng-Chang Tsai, Yun-Hsiu Tseng, and Wei-Fang Su

1 Introduction

There is a growing trend that millions of people suffer from glaucoma which is one of the neurodegenerative diseases [1]. The degeneration of optic nerves has limited regenerative capacity due to myelin and glial scar formation from lesion [2]. It is important to establish an efficient method to cure the optic nerve dysfunction. Neural tissue engineering is regarded as a hopeful method to transplant the cell containing a functional scaffold and restore the neural tissue function [3]. The optic nerve exhibits anisotropic hierarchical morphology to transmit the directional action potential signal [4]. Among the artificial biomimetic scaffolds used for neural tissue engineering, hydrogel is a preferred soft material because the 3D structure and water content are similar to the natural extracellular matrix (ECM) [5]. Also, the anisotropic structure is essential to the optic nerve due to the guidance of the neuron. Many methods have been developed to fabricate hydrogel with aligned structures such as self-assembled peptide and polymer under the magnetic field. For example, self-assembled peptide could be synthesized via solid-phase peptide synthesis with a precise peptide sequence [6–9]. However, the process is time-consuming, low yield, and high cost which limits the large-scale production. Then, the hydrogel made of magnetic responsive polymer combined with an external magnetic field is a non-invasive method [10–13]. However, the synthesis of polymer is complicated and the alignment of molecules using the cumbersome magnetic generator is not readily accessible. Herein, we propose to prepare self-assembly nanocomposite to obtain hydrogel with aligned structure. We first synthesized water-soluble

C.-C. Tsai · Y.-H. Tseng · W.-F. Su (✉)
Department of Materials Science and Engineering, National Taiwan University, Taipei, Taiwan
e-mail: suwf@ntu.edu.tw

W.-F. Su
Molecular Imaging Center, National Taiwan University, Taipei, Taiwan

K.-P. Lin et al. (eds.), *Future Trends and Challenges of Molecular Imaging and AI Innovation*, Springer Proceedings in Physics 272,
https://doi.org/10.1007/978-3-030-92786-8_1

polypeptide containing the neurotransmitter of glutamate, named sodium salt of poly (γ-benzyl-L-glutamate)$_{40}$-r-poly (L-glutamic acid)$_{60}$ (PBGA60-Na). Then, it was used as a crosslinker to gel the cationic cellulose nanofibrils (CNF^+) to form nanocomposite hydrogel through static charge and hydrogen bonding. The hydrogel with anisotropic structure was fabricated under a shear force. The hydrogel prepared without shear force to exhibit isotropic structure served as a control in this study. The aligned morphology of hydrogel could be observed by polarized optical microscope, scanning electron microscope, and X-ray tomography analyses at different scales of size. Owing to the high transmission capacity and spatial resolution of synchrotron radiation, the X-ray tomography could be reconstructed the 3D structure of material in situ. As a result, this study describes the first example of using the X-ray tomography to precisely observe the hydrogel with the aligned structure, which could provide the useful information needed in the clinic application.

2 Materials and Methods

A. Materials

Chemicals used for synthesis including L-glutamic acid γ-benzyl ester (Sigma-Aldrich), triphosgene (Sigma-Aldrich), benzylamine (Alfa Aesar), dichloroacetic acid (DCA, Acros), 33 wt% hydrogen bromide (HBr) in acetic acid (Acros), sodium bicarbonate ($NaHCO_3$, Fisher Scientific), ethyl acetate (EA, Macron), dimethylformamide (DMF, Sigma-Aldrich), tetrahydrofuran (THF, Macron), diethyl ether (J. T. Baker), methanol (Macron), and hexane (Macron). Chemicals for instrument analysis included ruthenium tetroxide, 0.5% stabilized aqueous solution (RuO4, Polysciences Inc.), chloroform-d (Merck), and trifluoroacetic acid-d (Acros). Cationic cellulose nanofibrils were purchased from Cellulose Lab.

B. Nomenclature

The abbreviation of the polymers was shown as follows: poly (γ-benzyl-L-glutamate) (PBG), poly (α-L-glutamic acid) with 60 mol% of hydrolysis (PBGA60), sodium salt of PBGA60 (PBGA60-Na), and cationic cellulose nanofibrils (CNF^+). The crosslinker (PBGA60-Na) and CNF^+ were abbreviated as C and M, respectively. The prefix *iso-* represented the isotropic morphology of hydrogel. The anisotropic hydrogel and isotropic hydrogel with a composition of 1.0 wt% CNF^+ and 0.25 wt% PBGA60-Na were named as M(1.0)C(0.25) and iso-M(1.0)C(0.25), respectively.

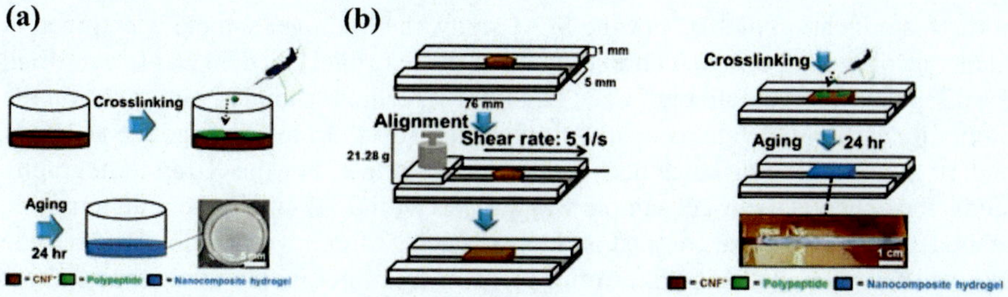

Fig. 1 Schematic diagram and actual sample diagram of **a** isotropic and **b** anisotropic hydrogel

C. Synthesis of PBGA60-Na

The PBGA60-Na was first prepared by a partial hydrolysis of PBG [14], and then a weak base treatment. The PBG (1 g) was dissolved in DCA (25 ml) first, and then adding HBr in acetic acid (33 wt%, 0.76 ml). After reacting at 31 °C for 2 h, the reaction solution was precipitated in cold ether, redissolved in THF, and recovered through precipitation in ether. The PBGA60 was purified five times from ether/THF cosolvent to obtain pure white solid, which was vacuum dried at 40 °C overnight (yield: 70%). The PBGA60 (0.2 g) was dissolved in $NaHCO_3$ solution (1 M, 6 ml) in a 25 ml two neck bottle. The solution was placed in dialysis membrane (MWCO 3500; Ordial D35; Orange Scientific) with deionized water to purify the polymer. The resulting product was freeze-dried at -53 °C by lyophilizer for two days to obtain a sponge-like white solid (yield: 88%).

D. Hydrogel preparation

For POM and SEM study, 100 μl of 1.0 wt% CNF^+ solution was placed in a 24 well plate first and then 100 μl of 0.25 wt% PBGA60-Na was added to the CNF^+ solution dropwise. The mixed solution was aged for 24 h to ensure complete gelation to prepare the isotropic hydrogel (iso-M(1.0)C(0.25)) by drop-casting as Fig. 1a shown. The anisotropic hydrogel (M(1.0)C(0.25)) was prepared by shear force shown as Fig. 1b. Two cover glasses (thickness ~ 1 mm) were pasted on each side of the glass substrate with a 5 mm wide groove between. 100 μl of 1.0 wt% CNF^+ solution was added in the groove. Shear the CNF^+ solution at 5 1/s shear rate under 21.28 g loading (Teflon piece and weight) to align CNF^+. 100 μl of 0.25 wt% PBGA60-Na was then added to the CNF^+ aligned solution dropwise throughout the CNF^+ solution. The mixed solution was aged for 24 h to ensure complete gelation. For the X-ray tomography study, only the halved volume (50 μl) of both PBGA60-Na and CNF^+ solution was needed for isotropic and anisotropic hydrogels.

E. Characterization of hydrogel morphology

The morphology of hydrogels at different scales was characterized using polarized optical microscope (POM, DM 2500 M, Leica), scanning electron microscope (SEM, JSM-6510, JEOL), and transmission X-ray microscope (TXM, NSRRC, Taiwan). For the POM study, the aqueous hydrogel sample was observed using cross-polar mode at the magnification of 10x. For the SEM study, the hydrogel sample was frozen in liquid nitrogen for 1 min to fix the structure of the hydrogel and then was freeze-dried for three days to maintain the frozen structure. The dried samples were fixed on the holder by carbon tape and coated with platinum (60 s, 20 mA) to operate at 10 kV and 10 mm working distance between lens and sample. For the X-ray tomography study, the aqueous hydrogel sample was stained with 0.1% w/v ruthenium tetroxide in ddH_2O to enhance the contrast of the hydrogels. After 3 h of staining, the sample was washed with methanol. Tear off a piece of hydrogel with a thickness of about 14 ~ 17 μm to fix on the holder by polyamide tape. 20 μl gold particle solution (400 ~ 500 nm) was added to the hydrogel as the target of the rotation. The sample was rotated from -75° to 75°. The 3D structure was reconstructed by computer simulation through 151 cross-section images taken per degree by TXM.

F. Statistical analysis of the alignment of the hydrogel

The azimuth angle of each fibril was counted in the same direction to get the angle distribution from -90° to 90° by ImageJ. The average azimuth angle was redefined as 0° to represent the direction of shear force. The alignment of anisotropic hydrogels can be determined by the azimuth angle distribution of the cellulose fibrils in POM images (n = 50) and TXM cross-section images (n = 10 at -75°, -50°, -25°, 0°, 25°, 50°, 75°).

3 Results

The morphology of the isotropic (iso-M(1.0)C(0.25)) and anisotropic (M(1.0)C(0.25)) hydrogels under the same composition is observed by POM, SEM, and X-ray tomography at different scale. From POM images shown in Fig. 2a, the fibrils of the anisotropic hydrogel have more consistent birefringence (blue) and arrange in nearly the same direction as the shear force compared with the isotropic hydrogel. The alignment of hydrogels is defined by the percentage of the azimuth angle from -5° to 5°. The statistical results of the azimuth angle distribution are shown in Fig. 2d. The alignment of the M(1.0)C(0.25) is 42.0%, while the iso-M(1.0)C(0.25) has no obvious directionality. The SEM images of the iso-M(1.0)C(0.25) and M(1.0)C(0.25) in Fig. 2b show the surface morphology of the dried hydrogels. The surface morphology is much more aligned after the shear force treatment. From the simulation results of the X-ray tomography shown in Fig. 2c, the fibrils arrangement of the isotropic hydrogel is disorder, while the direction of the shear force can be easily judged by the naked eye. The statistical alignment of the M(1.0)C(0.25) from the X-ray tomography in Fig. 2e is 23.3%.

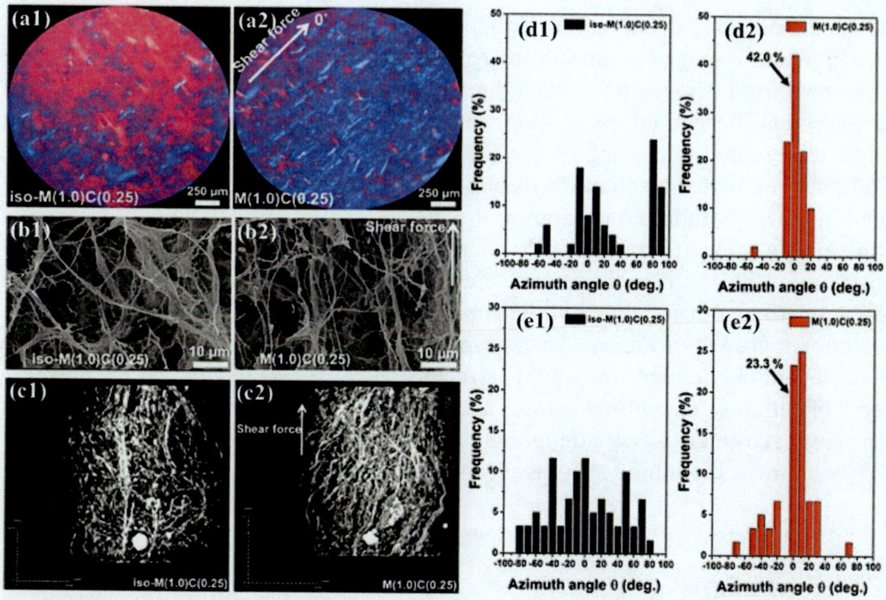

Fig. 2 The **a** POM **b** SEM **c** the simulation results of X-ray tomography images of (1) iso-M(1.0)C(0.25) and (2) M(1.0)C(0.25) and the statistical results from **d** POM and **e** TXM cross-section images

4 Discussion

The morphology of the biological scaffold is the critical factor in cell growth, including proliferation, differentiation, and migration [15, 16]. For neurons, the scaffold with specific directionality is a must to achieve the directional transmission of nerve signals. This study proposes an effective and low-cost method to prepare hydrogels with specific alignment. The shear force applied to the CNF^+ solution shown in Fig. 1b successfully aligns the cellulose along the direction of the shear force. Due to the optically positive order structure of the fibrils, the POM technique is an appropriate method to analyze the distribution of the fibrils in hydrogels at the microscale. The hydrogel without applying shear force shows the random direction distribution of fibrils. Neither blue (1st and 3rd quadrant) nor orange (2nd and 4th quadrant) part dominates in Fig. 2a1, while the blue part dominates after applying the shear force shows in Fig. 2a2. The SEM technique shows the surface morphology from micro to nanoscale as Fig. 2b shown, however, freeze-dried samples are required under the limitation of a high vacuum. Freeze-dried samples may distort the high-water-content hydrogel. Therefore the X-ray tomography becomes a powerful technique to study the 3D structure in situ at the sub-micro scale. Through the TXM cross-section images at different degrees, the real 3D structure of aqueous hydrogels can be reconstructed as Fig. 2c shown. Performing statistical analysis on POM and TXM cross-section images, the alignment of the anisotropic hydrogel, M(1.0)C(0.25), can be quantified as 42.0% as cellulose bundles arranged at the microscale and 23.3% as nanofibrils arranged at the sub-micro scale as Fig. 2d, e shown respectively. The alignment from TXM is lower than the one from POM. Since the result from TXM analyzed at different degree considers not only the azimuth angle but the elevation angle represented to the angle between the fibrils and the rotation axis. However, POM images are the result of fibrils projecting on the imaging plane which simplify the alignment without consideration the elevation angle effect causes the overestimation. In previous studies [17], micro-CT was widely used to identify the 3D structure of hydrogels at the microscale. Since the scale of neurons may be low to a few microns, the alignment at the sub-micro or nanoscale is desired to guide the directional growth of nerves. The sub-micro X-ray tomography used in this study helps us not only understand the 3D structure of hydrogels in situ, but the arrangement of cellulose nanofibrils similar to the nerves scale was successfully observed. The novel sub-micro X-ray tomography is the potential technique to identify the true 3D structure of biominic scaffolds for tissue engineering in the coming future.

5 Conclusions

The polypeptide with glutamate as a neurotransmitter has been proven to stimulate the growth and proliferation of neurons. This study successfully modified PBG to water-soluble polyelectrolyte, PBGA60-Na. The easily arranged cellulose nanofibrils (CNF^+) with good strength to support nerve growth can be quickly crosslinked into nanocomposite hydrogels through electrostatic force and hydrogen bonding force. An effective, low-cost method to align hydrogel along the direction of the shear force has been developed. The alignment of the cellulose bundles of aqueous hydrogels at the microscale observed by POM is up to 42.0%. To solve the distortion of freeze-dried samples, we chose the novel sub-micro X-ray tomography instead of SEM to observe the true 3D structure of aqueous hydrogels. The alignment of nanofibrils can reach 23.3% in the three-dimension. The novel sub-micro X-ray tomography is the potential technique to identify the true 3D structure of biominic scaffold to tissue engineering in the coming future.

Acknowledgements This research was supported financially by the Ministry of Science and Technology (MOST), Taiwan, under Contracts MOST 108–2221–E–002–027–MY3.

Conflict of Interest The authors declare that they have no conflict of interest.

References

1. Tham YC, Li X, Wong TY, Quigley HA, Aung T, Cheng CY (2014) Global prevalence of

glaucoma and projections of glaucoma burden through 2040: a systematic review and meta-analysis. Ophthalmology 121:2081–2090
2. Gokoffski KK, Peng M, Alas B, Lam P (2020) Neuro-protection and neuro-regeneration of the optic nerve: recent advances and future directions. Curr Opin Neurol 33:93–105
3. Amani H, Arzaghi H, Bayandori M, Dezfuli AS, Pazoki-Toroudi H, Shafiee A, Moradi L (2019) Controlling cell behavior through the design of biomaterial surfaces: a focus on surface modification techniques. Adv Mater Interfaces 6:1900572
4. Vrabec JP, Levin LA (2007) The neurobiology of cell death in glaucoma. Eye 21:S11–S14
5. Chimene D, Kaunas R, Gaharwar AK (2020) Hydrogel bioink reinforcement for additive manufacturing: a focused review of emerging strategies. Adv Mater 32:1902026
6. Sur S, Newcomb CJ, Webber MJ, Stupp SI (2013) Tuning supramolecular mechanics to guide neuron development. Biomaterials 34:4749–4757
7. Godbe JM, Freeman R, Burbulla LF, Lewis J, Krainc D, Stupp SI (2020) Gelator length precisely tunes supramolecular hydrogel stiffness and neuronal phenotype in 3D culture. ACS Biomater Sci Eng 6:1196–1207
8. Li A, Hokugo A, Yalom A, Berns EJ, Stephanopoulos N, McClendon MT, Segovia LA, Spigelman I, Stupp SI, Jarrahy R (2014) A bioengineered peripheral nerve construct using aligned peptide amphiphile nanofibers. Biomaterials 35:8780–8790
9. Berns EJ, Sur S, Pan L, Goldberger JE, Suresh S, Zhang S, Kessler JA, Stupp SI (2014) Aligned neurite outgrowth and directed cell migration in self-assembled monodomain gels. Biomaterials 35:185–195
10. Antman-Passig M, Shefi O (2016) Remote magnetic orientation of 3D collagen hydrogels for directed neuronal regeneration. Nano Lett 16:2567–2573
11. Omidinia-Anarkoli A, Boesveld S, Tuvshindorj U, Rose JC, Haraszti T, Laporte LD (2017) An injectable hybrid hydrogel with oriented short fibers induces unidirectional growth of functional nerve cells. Small 13:1702207
12. Rose JC, Gehlen DB, Haraszti T, Köhler J, Licht CJ, Laporte LD (2018) Biofunctionalized aligned microgels provide 3D cell guidance to mimic complex tissue matrices. Biomaterials 163:128–141
13. Rose JC, Cámara-Torres M, Rahimi K, Köhler J, Möller M, Laporte LD (2017) Nerve cells decide to orient inside an injectable hydrogel with minimal structural guidance. Nano Lett 17:3782–3791
14. Lin CY, Luo SC, Yu JS, Chen Ta, Su WF (2019) Peptide-based polyelectrolyte promotes directional and long neurite outgrowth. ACS Appl. Bio Mater, 518–526
15. Huang G, Li F, Zhao X, Ma Y, Li Y, Lin M, Jin G, Lu TJ, Genin GM, Xu F (2017) Functional and biomimetic materials for engineering of the three-dimensional cell microenvironment. Chem Rev 117:12764–12850
16. Li Y, Rodrigues J, Tomás H (2012) Injectable and biodegradable hydrogels: gelation, biodegradation and biomedical applications. Chem Soc Rev 41:2193–2221
17. Cengiz IF, Oliveira JM, Reis RL (2018) Micro-CT – a digital 3D microstructural voyage into scaffolds: a systematic review of the reported methods and results. Biomater Res 22:26

^{111}In-radiolabeling of Mesenchymal Stem Cell-derived Extracellular Vesicles for Quantitative In Vivo Imaging

Cheng-Hsiu Lu, Yi-An Chen, Chien-Chih Ke, Chao-Cheng Chen, and Ren-Shyan Liu

1 Introduction

Mesenchymal Stem Cells (MSCs) are multipotent stromal cells which can be found in different tissues including bone marrow, fat, cord blood and umbilical cord. MSCs are capable of self-renewal and have a great potential to differentiate into adipocytes, osteocytes, chondrocytes…, etc. [1–3]. Upon 2019, there were almost 920 clinical trials of using MSCs in the treatment of several types of disease, expounding that treatment of MSCs have been considered as a promising therapeutic strategy. MSCs exert their therapeutic effect by different mechanisms including immunomodulation, differentiation potential, homing and paracrine effect [4, 5]. Nevertheless, paracrine effect has been reported to be the major role in treatment of MSCs [6].

C.-H. Lu · R.-S. Liu (✉)
Program of Biomedical Science and Engineering, National Yang Ming Chiao Tung University, Taipei, Taiwan

Y.-A. Chen · R.-S. Liu
Institute of Clinical Medicine, National Yang Ming Chiao Tung University, Taipei, Taiwan

C.-C. Chen · R.-S. Liu
Department of Biomedical Imaging and Radiological Sciences, National Yang Ming Chiao Tung University, Taipei, Taiwan

Y.-A. Chen · C.-C. Ke · R.-S. Liu
Molecular and Genetic Imaging Core, Mouse Clinic, National Comprehensive Mouse Phenotyping and Drug Testing Center, Taipei, Taiwan

C.-C. Ke
Department of Medical Imaging and Radiological Sciences, Kaohsiung Medical University, Kaohsiung, Taiwan

R.-S. Liu
Department of Nuclear Medicine, Cheng Hsin Hospital, Taipei, Taiwan

K.-P. Lin et al. (eds.), *Future Trends and Challenges of Molecular Imaging and AI Innovation*, Springer Proceedings in Physics 272,
https://doi.org/10.1007/978-3-030-92786-8_2

Extracellular vesicles (EVs) are naturally generated vesicles from cells. These vesicles mediate the cell-to-cell communication, and are capable of horizontal transfer of protein, nucleic acids as well as lipids from parental cells to recipient cells (Samanta et al. in Acta Pharmacol Sin 39:501–513, 2018). Based on these inherent characteristics, EVs-based remedy has been considered as a promising therapeutic candidate to treat variety of diseases (Yu et al. in Int J Mol Sci 15:4142–4157, 2014). Therapeutic effects of MSC-derived EVs (MSC-EVs) has been demonstrated on injury repair, regulation of immune system and inhibition of tumorigenesis (Maji et al. in Am J Physiol Gastrointest Liver Physiol 312:G194–G200, 2017). Therefore, knowing the pharmacokinetics biodistribution and the fate of EVs are worthy for pursuit.

As a potential future medical regimen for human disease, the information of EVs' behavior in the living subjects after administration needs to be urgently gathered. However, an effective method to monitor the *in vivo* behavior of EVs has not yet been developed. Despite that several approaches have tried to localize the distribution of EVs *in vivo*, including optical imaging with luminescent or fluorescent probes, most of the cases have limitations of signal penetration (Rocco et al. in Stem Cells Int 2016:5029619, 2016).

Herein, we have developed a simple and valid approach by combination of radiolabeled EVs and SPECT/CT imaging to access the *in vivo* biodistribution of EVs in small animal. ^{111}In-oxine is a commercially available clinical imaging agent for the use of radiolabeling leukocytes, lymphocytes, NK cells, MSCs as well as liposomes (Kraitchman et al. in Circulation 112:1451–1461, 2005;Malviya et al. in Mol Pharm 13:1329–1338, 2016;). Due to its lipophilic property, ^{111}In-oxine is easily able to penetrate the cell membrane, into the cells, then the complex disassembles and ^{111}indium is trapped in the cytoplasm. In addition, the 2.8 days half-life of ^{111}Indium allows a prolonged observation period to monitor dynamic behavior of MSC-EVs. In this study, we established a platform by radiolabeling MSC-EVs with ^{111}In-oxine for SPECT/CT imaging, which has the potential for MSC-EVs trafficking *in vivo.*

2 Materials and Methods

A. Preparation of EVs

Conditioned medium was collected and MSC-derived EVs were isolated using Exo-Prep (Lonza) according to the manufacturer's protocol. In brief, conditioned medium was mixed with equal volume of Exo-Prep reagent and incubated for 60 min. The mixture was centrifuged at 10,000 g for 60 min and the supernatant was aspirated. The residual solution was centrifuged at 10,000 g for 5 min, and polymeric materials were subsequently eliminated using Sephadex G-25 in PD-10 desalting columns (GE Healthcare) [13]. After concentration to appropriate volume with 100 kDa Amicon® Ultra-15 Centrifugal Filter Units, the EVs were stored at -80 °C. To isolate EVs by sequential centrifugation, the conditioned medium was centrifugated at 10,000 g for

30 min at 4 °C to remove cell debris, and the supernatant was then ultracentrifuged at 100,000 g for 70 min at 4 °C using SW 28 rotor (Beckman Coulter). The pellets were resuspended in 1 mL PBS and ultracentrifuged at 100,000 g for 70 min at 4 °C [14].

B. Preparation and Purification of ^{111}In-EVs

15 μl of 68 mM oxine (8-hydroxyquinoline) in ethanol was mixed with 10 μl of ^{111}In (Indium chloride in 0.05 M HCl) in 400 μl of 0.1 M sodium acetate buffer (pH 5.5) and then incubated at 50 °C for 30 min. The lipophilic components were extracted three times with 0.1 ml of chloroform and then dried using a rotary evaporator. The extracted ^{111}In-oxine in chloroform was evaporated to dryness. This was followed by the addition of 20 μl of ethanol and 180 μl of labeling buffer containing EVs into the vial for one hour at 37 °C. Each mL of the labeling buffer contains 100 μg polysorbate 80, and 6 mg of HEPES (N-2-hydroxyethyl-piperazine-N'-2-ethane sulfonic acid) buffer in 0.75% sodium chloride solution. The pH range of labeling buffer was restricted to 6.5 ~ 7.5. The radioactivity of ^{111}In-oxine was measured using either a dose calibrator (CRC-15R, Capintec, Bioscan) or a gamma scintillation counter (Cobra II Autogamma, Packard).

3 Results

A. *Radiolabeling of WJ-MSC-EVs using ^{111} In-oxine*

Fraction 7–10, of which the radioactivity and protein were detected simultaneously, were considered as ^{111}In-oxine-EVs (Fig. 1a). The labeling efficiency was about 5%. ^{111}In-oxine-EVs incubated in the serum remained stable (85%) for at least 24 h (Fig. 1b), and the radiochemical purity of ^{111}In-oxine-EVs was higher than 90% as determined by the instant thin-layer chromatography (ITLC) method (Fig. 1c). In order to examine whether labeling procedure alter the characteristics of EVs, expression of surface markers and particle size of labeled EVs were assessed. The results showed that the size distribution of ^{111}In-oxine-EVs was similar to the unlabeled EVs and the peak of size distribution did not change markedly (Fig. 1d). Also, labeled EVs had similar expression of CD63 and HSP70 compared with those in parental EVs (Fig. 1e). In addition, to precisely calculate the dosage of EVs in radiolabeling process, the protein recovery rate was measured using PD-10 desalting columns followed by 100 kDa Amicon® Ultra-15 Centrifugal Filter Units. The result showed that PD-10 and 100 k filter retained 27% of EVs protein after labeling (Fig. 1f). These results suggested that labeling with ^{111}In-oxine did not alter the characteristics of WJ-MSC-EVs.

B. In vivo *imaging of WJ-MSC-EVs*

For in vivo imaging of WJ-MSC-EVs, 1.443 ± 0.13 MBq μCi ^{111}In-oxine-EVs were intravenously administrated into C57BL/6 mice followed by μSPECT/CT imaging

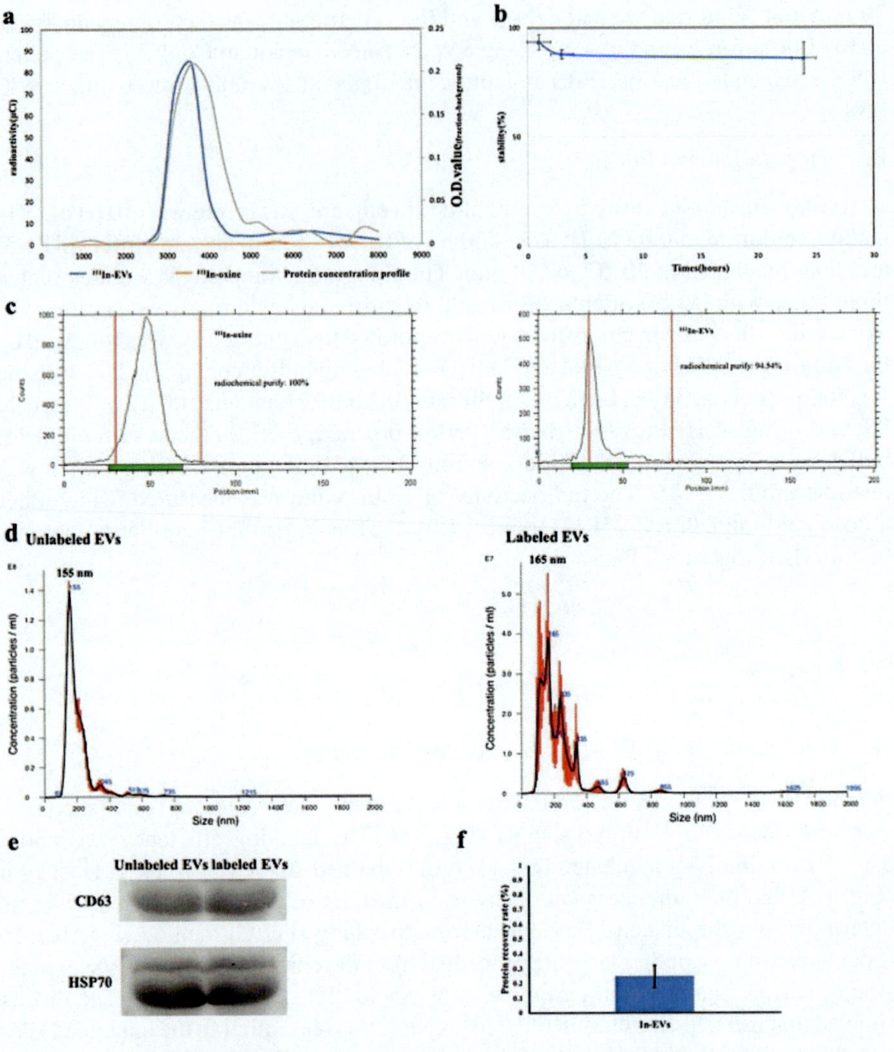

Fig. 1 Purification and characterization of ^{111}In-oxine labeled EVs. **a** Size exclusion was adopted to purify the ^{111}In-oxine labeled EVs through eluting PBS in PD-10 columns. The labeling efficiency was about 5%. **b** Stability of ^{111}In-oxine-EVs was examined in FBS. **c** ^{111}In-oxine and ^{111}In-oxine-EVs were analyzed by instant thin layered chromatography (ITLC) performing on a silica-gel-impregnated glass fiber sheet. The radiochemical purity of ^{111}In-oxine-EVs was general more than 90%. **d** The size distribution of ^{111}In-oxine-EVs was measured using NTA. **e** Representative western blot of labeled EVs and unlabeled EVs immunolabeled using CD63 and HSP70 antibody and visualized by ECL is shown. **f** Protein recovery rate was calculated by Bradford protein assay

at 1, 3, 6 and 24 h post-injection. A parallel imaging study was performed with ^{111}In-oxine, without labeling with EVs. One hour after the injection, ^{111}In-oxine-EVs mainly accumulated in the liver and spleen while most ^{111}In-oxine accumulated in the lung and kidney. The accumulated activity decreased gradually with time (Fig. 2a, b).

C. Ex vivo *biodistribution of WJ-MSC-EVs*

To further characterize the biodistribution, the mice were intravenously administrated ^{111}In-oxine-EVs or ^{111}In-oxine. After 1, 3, 6, 24 h, the mice were sacrificed to investigate biodistribution with ex vivo quantitation of radioactivity. Among all organs, the spleen, liver and kidney showed higher radioactivity in mice at one hour post iv injection of ^{111}In-oxine-EVs. The radioactivity accumulation in the liver, spleen and kidney gradually declined until 24 h post-injection (Fig. 3a). The radioactivity

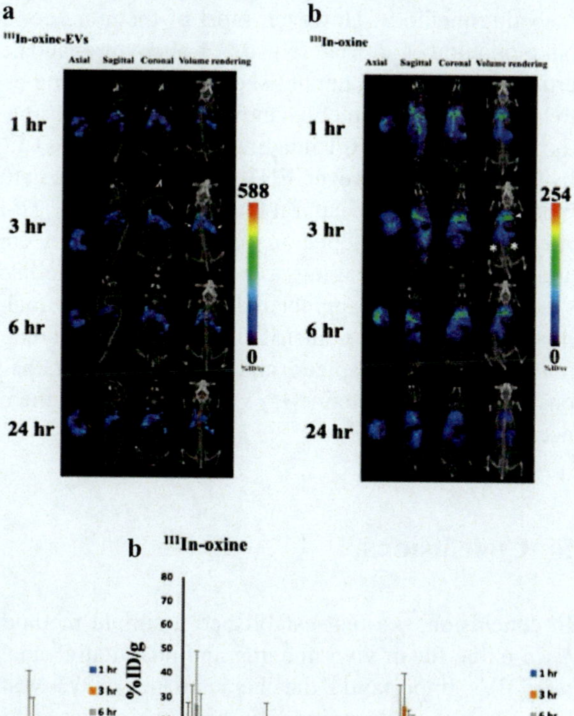

Fig. 2 In vivo μSPECT/CT imaging of ^{111}In-oxine-EVs. **a** After intravenous injection of ^{111}In-oxine-EVs, μSPECT/CT images was acquired at 1 h, 3 h, and 6 h as well as 24 h in C57BL/6 mice. **b** Micro-SPECT/CT images of ^{111}In-oxine administered groups. White arrow, spleen; dotted white arrow, liver; white arrow head, lung; white asterisk, kidney

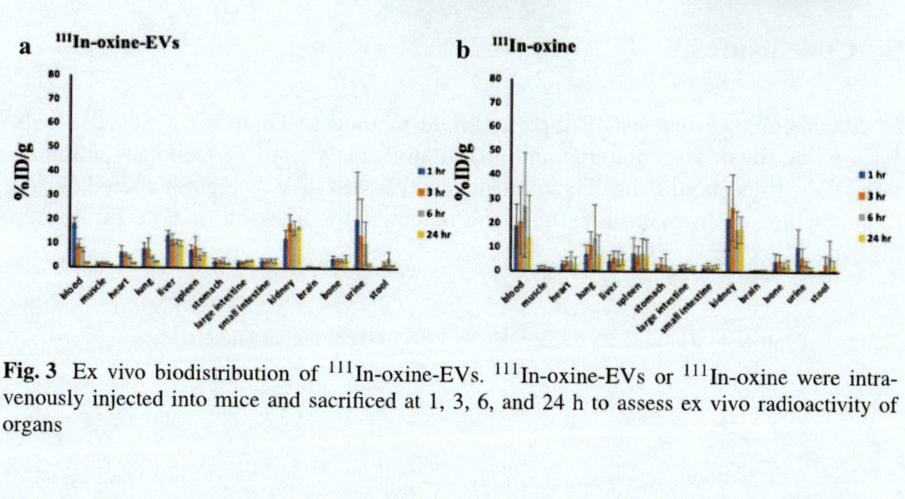

Fig. 3 Ex vivo biodistribution of ^{111}In-oxine-EVs. ^{111}In-oxine-EVs or ^{111}In-oxine were intravenously injected into mice and sacrificed at 1, 3, 6, and 24 h to assess ex vivo radioactivity of organs

in urine was much higher than that in stool indicating that the metabolites of ^{111}In-oxine-EVs were mainly eliminated via renal excretion. The different biodistribution profiles were observed specifically in the blood, lung and kidney in ^{111}In-oxine administered group (Fig. 3b).

4 Discussion

The present study demonstrated that WJ-MSC-EVs were able to be directly and simply labeled with ^{111}In-oxine, and retained the characteristics of MSC-EVs after radiolabeling. Moreover, in vivo μSPECT/CT imaging analysis and ex vivo quantitation demonstrated the exogenous WJ-MSC-EVs labeled with ^{111}In-oxine accumulated in the liver, spleen and kidney.

Several approaches have been established to monitor EVs in vivo using radiolabeling methods. However, most of them are complicated and time-consuming. Nikolopoulou et al. tried to use ^{131}I-phenylpropanoate to label EVs in 2015, but the article still remained unpublished. Besides, Hwang et al. developed a direct method to label exosome-mimetic nanovesicles (ENVs) using ^{99m}Tc-HMPAO. In that study, the in vivo μSPECT/CT imaging of ^{99m}Tc-HMPAO-ENVs highly accumulated in the liver and spleen. However, ENVs as alternatives to natural EVs, may not represent the biodistribution of the natural EVs labeled with ^{99m}Tc-HMPAO-ENVs. In the current study, we have developed an optimal approach by combination of ^{111}In-oxine-EVs and μSPECT/CT imaging, to demonstrate the biodistribution of WJ-MSC-EVs.

^{111}In-oxine is an appreciated potent agent for tracking of MSC-EVs. The *in vivo* imaging and *ex vivo* biodistribution demonstrated that ^{111}In-oxine-EVs clearly accumulated in the liver, spleen, and kidney. However, the safety and toxicity of continuously injected ^{111}In-oxine-EVs is still unknown, the more comprehensive studies is necessary.

5 Conclusions

In conclusion, we that established a simple method to label WJ-MSC-EVs with ^{111}In-oxine for in vivo imaging and quantitative analysis of exogenously administered EVs. Importantly, the characteristics of WJ-MSC-EVs were not altered during radiolabeling. ^{111}In-oxine-EVs have the potential for tracking of EVs by in vivo imaging.

Conflict of Interest The authors declare that they have no conflict of interest.

References

1. Lindner U, Kramer J, Rohwedel J, Schlenke P (2010) Mesenchymal stem or stromal cells: toward a better understanding of their biology? Transfus Med Hemother 37:75–83
2. Kern S, Eichler H, Stoeve J, Kluter H, Bieback K (2006) Comparative analysis of mesenchymal stem cells from bone marrow, umbilical cord blood, or adipose tissue. Stem Cells 24:1294–1301
3. Zeng R, Wang LW, Hu ZB et al. (2011) Differentiation of human bone marrow mesenchymal stem cells into neuron-like cells in vitro. Spine (Phila Pa 1976) 36:997–1005
4. Squillaro T, Peluso G, Galderisi U (2016) Clinical trials with mesenchymal stem cells: an update. Cell Transplant 25:829–848
5. Hoogduijn MJ, Lombardo E (2019) Mesenchymal stromal cells anno 2019: dawn of the therapeutic Era? Concise review. Stem Cells Transl Med 8:1126–1134
6. Linero I, Chaparro O (2014) Paracrine effect of mesenchymal stem cells derived from human adipose tissue in bone regeneration. PLoS One 9:e107001
7. Samanta S, Rajasingh S, Drosos N, Zhou Z, Dawn B, Rajasingh J (2018) Exosomes: new molecular targets of diseases. Acta Pharmacol Sin 39:501–513
8. Yu B, Zhang X, Li X (2014) Exosomes derived from mesenchymal stem cells. Int J Mol Sci 15:4142–4157
9. Maji S, Matsuda A, Yan IK, Parasramka M, Patel T (2017) Extracellular vesicles in liver diseases. Am J Physiol Gastrointest Liver Physiol 312:G194–G200
10. Di Rocco G, Baldari S, Toietta G (2016) Towards therapeutic delivery of extracellular vesicles: strategies for in vivo tracking and biodistribution analysis. Stem Cells Int 2016:5029619
11. Kraitchman DL, Tatsumi M, Gilson WD et al (2005) Dynamic imaging of allogeneic mesenchymal stem cells trafficking to myocardial infarction. Circulation 112:1451–1461
12. Malviya G, Nayak T, Gerdes C, Dierckx RA, Signore A, de Vries EF (2016) Isolation and (111)In-Oxine Labeling of Murine NK Cells for assessment of cell trafficking in orthotopic lung Tumor model. Mol Pharm 13:1329–1338
13. Li P, Kaslan M, Lee SH, Yao J, Gao Z (2017) Progress in exosome isolation techniques. Theranostics 7:789–804
14. Thery C, Amigorena S, Raposo G, Clayton A (2006) Isolation and characterization of exosomes from cell culture supernatants and biological fluids. Curr Protoc Cell Biol Chapter 3:Unit 3 22

A Comparative Study of Drug Distribution Using Photoacoustic Imaging and Radionuclide-Based Sampling in PEGylated Gold Nanostars-Mediated Photothermal Therapy

Wing Sum Lau, Chao-Cheng Chen, Ju-Yu Chen, Hui-Wen Chan, Deng-Yuan Chang, Hsin-Ell Wang, and Liang-Ting Lin

1 Introduction

Triple-negative breast cancer (TNBC) is aggressive with a poor prognosis and high relapse rate, contributing to 15–20% of overall breast carcinomas [1]. TNBC is defined by lacking estrogen receptors and progesterone receptors, as well as without overexpression of the human epidermal growth factor receptor 2 [2]. Conventional chemotherapy remains the mainstay of TNBC treatment regimen, which often conversely lead to systemic toxicity and chemoresistance [3]. Nanomedicine emerges as a novel therapeutic strategy to enhance the therapeutic efficacy and minimize side-effects [4]. Gold nanoparticles (AuNPs) are extensively studied as a theranostic agent for imaging and light-induced therapy, due to the tunable optical characteristics, simple surface modifications and biocompatibility [5]. AuNPs exhibit strong absorption of near-infrared (NIR) light. Upon laser irradiation, AuNPs convert the absorbed light into heat, which can be used in sensitizing tumor to photothermal therapy (PTT) [6]. Factors affecting the effectiveness of AuNPs-mediated PTT include the

W. S. Lau (✉) · J.-Y. Chen · L.-T. Lin
Department of Health Technology and Informatics, The Hong Kong Polytechnic University, Hung Hom, Hong Kong, China

L.-T. Lin
e-mail: lt.lin@polyu.edu.hk

C.-C. Chen · H.-W. Chan · D.-Y. Chang · H.-E. Wang
Department of Biomedical Imaging and Radiological Sciences, National Yang Ming Chiao Tung University, Taipei, Taiwan

K.-P. Lin et al. (eds.), *Future Trends and Challenges of Molecular Imaging and AI Innovation*, Springer Proceedings in Physics 272,
https://doi.org/10.1007/978-3-030-92786-8_3

photothermal conversion efficiency and AuNPs concentration within the irradiated area. Gold nanostars have a large surface-to-volume ratio and superior photothermal conversion efficiency, which can be a promising candidate for PTT [7]. Moreover, AuNPs tend to accumulate in tumor via enhanced permeability and retention (EPR) effect. Therefore, the in vivo distribution of photosensitizer needs to be carefully evaluated, in order to achieve tumor-selective PTT.

Molecular imaging plays a crucial role in guiding preclinical PTT studies by non-invasively tracking the photothermal agent accumulation along time. Photoacoustic imaging (PAI) is an advanced imaging modality combining the use of NIR light and ultrasound. AuNPs in tissue absorb light from the pulsed laser and generate acoustic waves by thermoelastic expansion [8]. Meanwhile, the changes in tumor microenvironment can be examined by analyzing PAI signals from oxyhemoglobin (HbO_2) and deoxyhemoglobin (Hb).

On the other hand, radiolabeling of AuNPs allows an investigation of the biodistribution and pharmacokinetics profile, conducted ex vivo or non-invasively using positron emission tomography (PET) and single photon emission computed tomography (SPECT), which are highly sensitive and extensively used in clinical applications. In the present work, we monitored the distribution of PEGylated gold nanostars (pAuNSs) using PAI and radionuclide-based biodistribution, thus determined the optimal treatment window for pAuNSs-mediated PTT. Afterwards, the therapeutic efficacy was evaluated and the potentials of theranostic pAuNSs for TNBC were discussed.

2 Methods & Materials

A. *Preparation of PEGylated Gold Nanostars*

The gold nanostars (AuNSs) were synthesized and modified as previously elaborated [9]. Briefly, AuNSs were formed by reducing hydrogen tetrachloroaurate solution with HEPES buffer, and followed by PEGylation to obtain pAuNSs. For radionuclide-based analysis, pAuNSs was tagged with indium-111 using DTPA as chelator, in order to acquire indium-111 labelled pAuNSs (^{111}In-DTPA-pAuNSs). The absorption spectrum of pAuNSs was validated by UV–visible spectrophotometer (Agilent 8453; Agilent Technologies, CA, USA). The labelling efficiency of radioactive surrogate was measured using the instant thin-layer chromatography (iTLC) with radio–thin-layer chromatography scanner (Bioscan AR2000; Bioscan, TriFoil Imaging Inc., CA, USA).

B. *Orthotopic Triple-Negative Breast Cancer Animal Model*

4T1_PB3R murine mammary tumor cell line was used for establishing the orthotopic TNBC model in female BALB/c nude mice (6 weeks old) [10]. Cells were cultured in RPMI-1640 medium with 10% fetal bovine serum and 1% penicillin–streptomycin solution, maintained in 37 °C, 5% CO2 incubator. 1×10^6 4T1_PB3R cells were harvested and suspended in 50 μl of phosphate buffered saline solution, then injected into the mammary pad. Animal weight and tumor size were measured and recorded every three days, until the tumor volume reached 100 mm^3 for in vivo studies. Animal experiments were in compliance with the ethical approval granted from the Animal Subjects Ethics Sub-Committee (ASESC) of The Hong Kong Polytechnic University (19–20/45-HTI-R-GRF).

III. In vivo *Biodistribution Study*

The distribution of ^{111}In-DTPA-pAuNSs (36.4 ± 0.8 μCi/mice) in the 4T1 tumor-bearing mice ($n = 2$) was investigated at 1, 4, 24, and 48 h post-intravenous injection. At each designated time point, two tumor-bearing mice was sacrificed, and the tissues/organs of interest were harvested, washed with water and weighted, followed by radioactivity measurement with radioactive decay correction using the Wallace 1470 Wizard Gamma counter (GMI, Inc., Ramsey, MN, USA). Data were expressed as the percentage injected dose per gram of tissues/organs (%ID/g).

IV. In vivo *Photoacoustic Imaging*

PAI scan was performed using Multispectral Optoacoustic Tomography (MSOT) inVision 128 system (iThera Medical, Germany). The tumor-bearing mice ($n = 2$) were anesthetized with 2% isoflurane and secured inside a designated animal holder, then placed inside the imaging chamber filled with 34 °C deionized water. Pre and post-injection MSOT scans of tumor region were acquired at a series of time intervals (1, 2, 4, 24, 48 h) upon pAuNSs administration via tail vein (10 mg Au/kg). MSOT images were reconstructed by model-based method and component signals were spectrally unmixed.

E. *Evaluation of Photothermal Therapy Efficacy*

For treatment, tumor-bearing mice were randomly divided into pAuNSs-PTT and laser-only group ($n = 3$), which received intravenous injection of pAuNSs (10 mg Au/kg) or normal saline respectively. The tumor was irradiated by 808 nm laser (1 W/cm^2) repeatedly for 3 min twice with a 2 min rest in between at 24 h post-injection. The surface temperature of tumor was recorded by infrared camera during irradiation (Fluke Corp., WA, USA). To assess the therapeutic efficacy of pAuNSs-mediated PTT, the tumor growth was monitored twice a week, with the volume determined by caliper measurement and calculated as volume = $(\text{width}^2 \times \text{length})/2$. The tumor burden was also evaluated using bioluminescent imaging upon the intraperitoneal injection of D-luciferin (150 mg/kg) (VivoGlo Luciferin; Promega Corp., WI, USA) and imaged with IVIS In Vivo Animal Imaging Systems (Perkin Elmer Inc., MA, USA).

3 Results and Discussion

A. *Synthesis of PEGylated Gold Nanostars*

The UV–visible-NIR spectrum in Fig. 1 demonstrated a strong absorption of pAuNSs within the first NIR window (650–950 nm), where the transparent window of least photon attenuation from Hb and HbO_2 locates [11]. The absorption peak was at 820 nm. The labelling efficiency of ^{111}In-DTPA-pAuNSs was greater than 85%.

B. In vivo *Biodistribution Study*

According to the biodistribution results reported in Table 1, ^{111}In-DTPA-pAuNSs eliminated rapidly from blood after receiving the intravenous injection. The radioactivity dropped from 21.51 ± 0.99%ID/g at 1 h post-injection (p.i.) to 0.48 ± 0.12%ID/g at 24 h p.i.. The reticuloendothelial system exhibited high uptake of ^{111}In-DTPA-pAuNSs within the experimental period. In addition, significant radioactivity in urine could be attributed to the ^{111}In-DTPA-pAuNSs and their radiolabeled PEG fragments. The highest tumor uptake was demonstrated at 4 h p.i. (2.08 ± 0.17%ID/g), remained plateau till 24 h p.i., and then dropped to 1.34 ± 0.12%ID/g at 48 h p.i.. The calculated concentration of pAuNSs in 4T1 tumor reached approximately 4 ppm at 4 h p.i.

III. *Detection of pAuNSs Photoacoustic Signals*

Pre-injection MSOT images of tumor were taken with baseline signals of pAuNSs, Hb and HbO_2 resolved (Fig. 2a). During 2–24 h p.i., pAuNSs signals in large vessels diminished and the tumor periphery was gradually enhanced, enabling a good differentiation of tumor vasculature (Fig. 2b). Heterogenous probe distribution in tumor was noticed, where pAuNSs predominantly accumulated in the perivascular spaces. Moreover, the quantification of pAuNSs signal intensities showed a gradual saturation during 2-24 h (Fig. 2c). MSOT signals in tumor reached a four-fold increment at 24 h (2.53 ± 0.71 a.u.) comparing to the baseline (0.56 ± 0.02 a.u.), which could

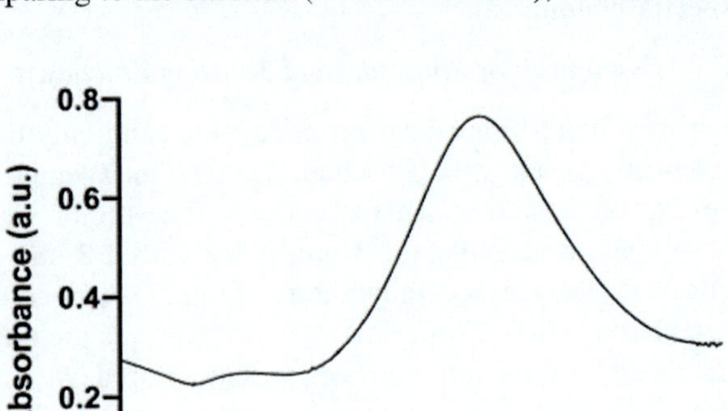

Fig. 1 Absorption spectrum of PEGylated gold nanostars (pAuNSs) characterized by UV–vis spectrophotometer

Table 1 Biodistribution of ^{111}In-DTPA-pAuNSs in the 4T1 tumor-bearing mouse model ($n = 2$)

	1 h	4 h	24 h	48 h
Blood	21.51 ± 0.99	5.45 ± 0.94	0.48 ± 0.12	0.13 ± 0.01
Liver	17.81 ± 1.35	21.36 ± 0.34	19.75 ± 0.18	26.43 ± 1.71
Spleen	34.49 ± 9.35	32.58 ± 5.57	22.97 ± 0.47	23.87 ± 1.08
Kidney	4.50 ± 0.10	4.64 ± 0.36	2.71 ± 0.16	2.26 ± 0.00
Muscle	0.30 ± 0.07	0.23 ± 0.04	0.19 ± 0.19	0.06 ± 0.00
Urine	135.77 ± 18.25	67.36 ± 23.83	3.81 ± 0.55	3.58 ± 0.90
Feces	0.03 ± 0.01	1.29 ± 0.43	0.57 ± 0.07	0.34 ± 0.11
Tumor	1.39 ± 0.23	2.08 ± 0.17	2.07 ± 1.21	1.34 ± 0.12

Data expressed as the percentage of the injected dose per gram of an organ (%ID/g)

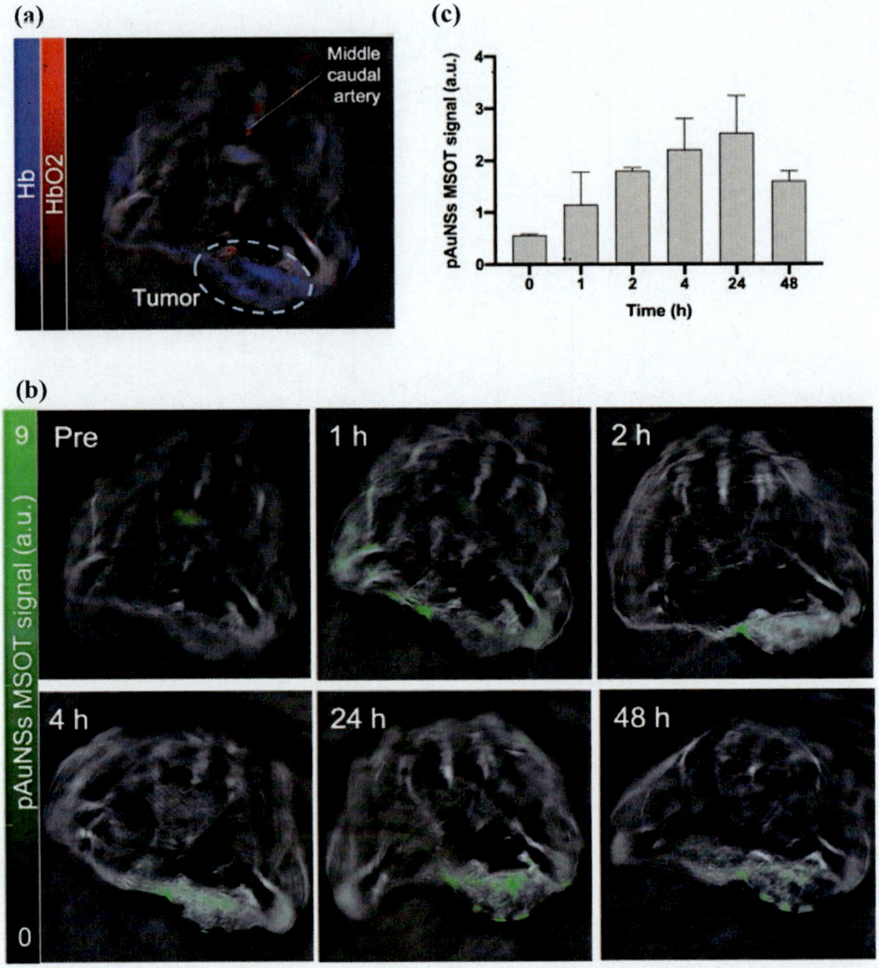

Fig. 2 In vivo MSOT tracking of pAuNSs uptake in 4T1 tumor. **a** Pre-injection image of the deoxyhemoglobin (blue) and oxyhemoglobin (red) distribution ($n = 2$). **b** Transverse images of tumor at different time intervals upon i.v. injection of pAuNSs (green) (10 mg Au/kg, green). **c** MSOT intensity of pAuNSs signals in tumor over time ($n = 2$)

be attributed to the EPR effect usually taking 6-24 h [12]. From the biodistribution analysis and MSOT images, most pAuNSs in bloodstream had eliminated at 24 h, explaining the MSOT signals at 24 h could be contributed by pAuNSs predominantly trapping in the tumor, suggesting the best treatment timepoint for tumor-selective PTT.

IV. *Evaluation of Photothermal Therapy Efficacy*

The surface temperature of the tumor region during laser exposure was recorded using infrared camera as shown in Fig. 3. The maximal tumor temperature after

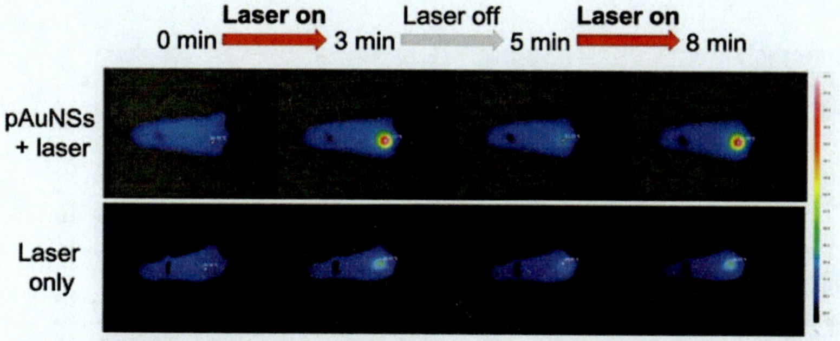

Fig. 3 In vivo photothermal conversion effect. Thermal images of 4T1 tumor-bearing mice receiving laser irradiation with administration of pAuNSs and normal saline respectively (n = 3)

808 nm laser irradiation reached 55.3 ± 2.88 °C in mice administered with pAuNSs, comparing to 36.8 ± 0.24 °C in the laser-only group. The tumor temperature rise repeatedly exceeded 50 °C, which could cause irreversible thermal ablation of the tumor tissue, thus inducing cellular death by necrosis [13].

Bioluminescent images revealed a significant reduction of luciferase activity in PTT-treated group, indicating suppression of the tumor viability (Fig. 4a). On the contrary, the laser-only group exhibited distant metastases on day 21 post-laser irradiation. Furthermore, the tumor burden became undetectable by caliper measurement in pAuNSs-mediated PTT group, until day 14 post-PTT where local recurrence occurred in one of the three mice (Fig. 4b). However, no sign of tumor shrinkage was noticed in the laser-only group, suggesting that laser alone did not have killing effect on 4T1 tumor. There was no apparent loss in body weight after treatment in both groups, indicating that there was no observable acute toxicity caused by pAuNSs-mediated PTT or laser irradiation (Fig. 4c). Both tumor volume and bioluminescent signals corroborated a remarkable tumor suppressive effect brought by pAuNSs-mediated PTT, whereas the mechanism of cellular damage required further investigation by immunohistochemistry studies.

4 Conclusion

Herein, we presented the photoacoustic imaging and radiolabeling technique to investigate the distribution of PEGylated gold nanostars in orthotopic 4T1-tumor bearing mice at different time intervals. Both methods revealed that the maximum accumulation of pAuNSs in tumor sustained within 2-24 h, ascribed to the EPR effect. MSOT images demonstrated a predominant accumulation in perivascular spaces, which could potentially be a treatment strategy of disrupting vessels focally. With the intrinsic optical property, pAuNSs could simultaneously act as a photoacoustic contrast and photothermal sensitizer. This work shed light on the theranostic options

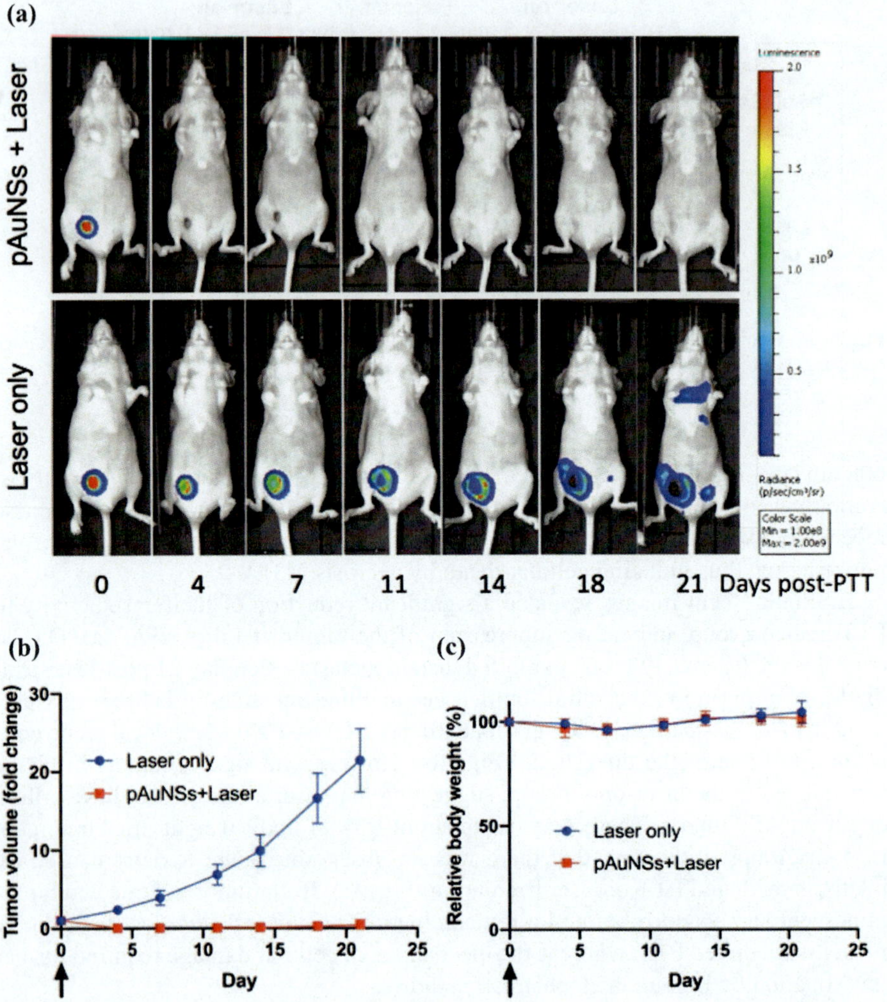

Fig. 4 Therapeutic efficacy and toxicity of pAuNSs-mediated photothermal therapy (PTT). **a** Bioluminescent images of orthotopic 4T1 tumor growth. Relative changes of tumor volume (**b**) and body weight (**c**). Tumor volume was measured by a caliper ($n = 3$). Data were represented as mean ± S.D. The black arrow indicates the initial time point of treatment

of using pAuNSs as a potential nanoplatform for image-guided cancer therapy, also unveiled the promising antitumor effect of pAuNSs-mediated PTT in orthotopic TNBC model.

Acknowledgements This work is supported by Early Career Scheme (25103719) by the Research Grant Council, HKSAR.

Conflicts of Interest The authors declare that there are no conflicts of interests regarding this study.

References

1. Jitariu A-A, Cîmpean AM, Ribatti D, Raica M (2017) Triple negative breast cancer: the kiss of death. Oncotarget 8:46652
2. Abramson VG, Lehmann BD, Ballinger TJ, Pietenpol JA (2015) Subtyping of triple-negative breast cancer: implications for therapy. Cancer 121:8–16
3. Gangi A, Mirocha J, Leong T, Giuliano AE (2014) Triple-negative breast cancer is not associated with increased likelihood of nodal metastases. Ann Surg Oncol 21:4098–4103
4. Thakur V, Kutty RV (2019) Recent advances in nanotheranostics for triple negative breast cancer treatment. J Exp Clin Cancer Res 38:1–22
5. Knights OB, McLaughlan JR (2018) Gold nanorods for light-based lung cancer theranostics. Int J Mol Sci 19:3318
6. Li W, Chen X (2015) Gold nanoparticles for photoacoustic imaging. Nanomedicine 10:299–320
7. Hao F, Nehl CL, Hafner JH, Nordlander P (2007) Plasmon resonances of a gold nanostar. Nano Lett 7:729–732
8. Valluru KS, Chinni BK, Rao NA, Bhatt S, Dogra VS (2009) Basics and clinical applications of photoacoustic imaging. Ultras Clin 4:403–429
9. Chen C-C, Chang D-Y, Li J-J et al (2020) Investigation of biodistribution and tissue penetration of PEGylated gold nanostars and their application for photothermal cancer treatment in tumor-bearing mice. J Mater Chem B 8:65–77
10. Chen Y-L, Wang C-Y, Yang F-Y et al (2014) Synergistic effects of glycated chitosan with high-intensity focused ultrasound on suppression of metastases in a syngeneic breast tumor model. Cell Death Dis 5:e1178–e1178
11. Ntziachristos V, Razansky D (2010) Molecular imaging by means of multispectral optoacoustic tomography (MSOT). Chem Rev 110:2783–2794
12. Maeda H, Nakamura H, Fang J (2013) The EPR effect for macromolecular drug delivery to solid tumors: improvement of tumor uptake, lowering of systemic toxicity, and distinct tumor imaging in vivo. Adv Drug Deliv Rev 65:71–79
13. Kaur P, Aliru ML, Chadha AS, Asea A, Krishnan S (2016) Hyperthermia using nanoparticles–promises and pitfalls. Int J Hyperth 32:76–88

CMR-Derived Extracellular Volume Fraction is Associated with the Ventricular Wall Thickness in Patients with Hypertrophic Cardiomyopathy

Shun-Chung Yang, Hung-Wen Chiu, and Mao-Yuan Su

1 Introduction

Hypertrophic cardiomyopathy (HCM) is a hereditary heart disease [1, 2]. The main cause is abnormal muscle alignment of the heart muscle [3–5], which leads to poor contraction of the heart and compensatory hypertrophy of the myocardium, resulting in thickening of the myocardium and the appearance of ventricular hypertrophy [6, 7]. This hypertrophy and lack of elasticity may lead to difficulty in ventricular relaxation and heart failure [1, 8, 9]. Left ventricular hypertrophy can develop due to factors such as hypertension, aortic stenosis or diffuse myocardial fibrosis (DMF) etc. [10–12]. Myocardial fibrosis refers to tissue damage and scarring caused by ischemia, inflammation and hypertension of the myocardium. Previous studies have demonstrated that HCM presents the DMF, which correlate with estimated LV filling pressure, suggesting a mechanistic link between DMF and abnormal diastolic function. However, it remains unknown whether the degree of DMF is associated with LVH in patients with HCM.

Cardiovascular magnetic resonance (CMR) has been widely used to quantify DMF by calculating the extracellular volume fraction (ECV) of the myocardium [13, 14].

S.-C. Yang (✉) · H.-W. Chiu
Graduate Institute of Biomedical Informatics, Taipei Medical University, Taipei, Taiwan

H.-W. Chiu
e-mail: hwchiu@tmu.edu.tw

S.-C. Yang · M.-Y. Su
Taipei Association of Radiological Technologists, Taipei, Taiwan
e-mail: maoyuansu@gmail.com

Department of Medical Imaging, National Taiwan University Hospital, Taipei, Taiwan

M.-Y. Su
Department of Medical Imaging and Radiological Technology, Yuanpei University of Medical, Technology, Hsinchu, Taiwan

K.-P. Lin et al. (eds.), *Future Trends and Challenges of Molecular Imaging and AI Innovation*, Springer Proceedings in Physics 272,
https://doi.org/10.1007/978-3-030-92786-8_4

In this study, our aim was to investigate whether the severity of LVH is associated with the degree of DMF in patients with HCM.

2 Material and Methods

A. Patient selection

This study is a retrospective analysis and the CMR was performed at the National Taiwan University Hospital. A total of 125 patients who had a clinical evidence of HCM and 58 patients with Brugada syndrome (Brs) which was considered as the relative normal heart control were enrolled in this study. Patient with Brs had abnormal electrical potential activity in the heart, resulting in arrhythmias, but showed no significant different in the structure and function with healthy subjects were treated as control group. All study participants provided written informed consent.

B. CMR sequences

We performed CMR using a clinical 1.5-T scanner (Area 1.5 T, Siemens, Erlangen, Germany). Cine CMR was performed using a segmented balanced steady-state gradient echo pulse sequence with a retrospective electrocardiography (ECG)-R-wave trigger. The scanning parameters were as follows: repetition time (TR), 3.0 ms; echo time (TE), 1.5 ms; matrix size, 256 × 208; and spatial resolution, 1.21 mm. Multiple short-axis (SA) slices were prescribed in the left ventricle (LV) from the mitral orifice to apex with slice thickness of 8 mm and gap of 2 mm. The true temporal resolution was 63 ms, and 30 cardiac phases were reconstructed retrospectively for each slice level.

Myocardial T1 mapping was performed with an electrocardiography (ECG)-triggered modified Look-Locker inversion recovery (MOLLI) sequence before and 10 min after 0.15 mmol/kg intravenous administration of the gadolinium-based contrast agent (Dotarem, Guerbet, France). The MOLLI protocol used 2 Look-Locker cycles to acquire 7 images over 11 heartbeats, and the scanning parameters were as follows: TR/TE, 1.9 ms/1.0 ms; minimum inversion time, 110 ms; inversion time increment, 80 ms; matrix size, 256 × 192; slice thickness, 6 mm; spatial resolution, 1.28 mm; number of inversions, 2; images acquired after first inversion, 5; pause, 4 heartbeats; and images acquired after second inversion, 2 [15]. Three evenly spaced SA slices were acquired from the LV base to the apex. After postcontrast T1 acquisition, late gadolinium enhancement (LGE) images were acquired using an ECG-triggered phase-sensitive inversion recovery prepared segmented fast gradient-echo pulse sequence to identify the focal fibrosis or scarring.

C. Image analysis

Cine images were used to quantify the LV function and mass by determination of the endocardial and epicardial borders for each diastolic and systolic frame on all SA slices and then normalized by body surface area (BSA). Quantitative analysis of

ECV and regional wall thickness (WT) were performed on T1 maps. The ECV was calculated from pre- and postcontrast T1 maps using a region-based method and was then calibrated with the last available hematocrit (Hct) data. Four different regions, including the anterior, septal, posterior and lateral wall regions of the LV, and the area in the central area of the LV cavity were drawn on the T1 map at the mid-ventricular slice. The average T1 values of the segmented regions of interest were then computed. After obtaining the pre- and postcontrast T1 values, ECV values were calculated by using the following formula [16]:

$$ECV = \frac{\frac{1}{T1_{post}^{myo}} - \frac{1}{T1_{native}^{myo}}}{\frac{1}{T1_{post}^{blood}} - \frac{1}{T1_{native}^{blood}}} * (1 - Hct)$$

The WTs of the same regions were measured at the same level of the T1 mapping. Group comparisons in LV function, ECV and WT were compared between two group. The correlation between ECV and WT, including averaged and each segment, were assess for each group. In addition, the regional ECV was excluded if the regional myocardium showed hyper-enhancement on the LGE image. Image analysis was performed using software developed in-house provided by MATLAB 7.9 (MathWorks, Inc., Natick, Massachusetts) and Image J (NIH, USA).

D. Statistical analysis

Data are expressed as mean ± standard deviation (SD). Group comparisons in the myocardial ECV, WT and LV functional indexes were tested by the nonparametric Mann–Whitney U test. Linear regression was used to obtain a correlation coefficient between regional ECV and WT for each group. The statistical tests were two-tailed, and statistical significance was defined as $p < 0.05$. Statistical analyses were performed using SPSS software package version 20.0 (SPSS Inc., Chicago, IL, USA).

3 Results

A. Patient characteristics and Group comparisons

The demographics of the study population are summarized in Table 1. The HCM group was older but body surface area was similar as compared with control group. There were significant differences in LV function and mass between the patient and control groups (Table 1). For the group comparison, HCM group had a significant higher averaged ECV compared to that in control group (27.5 ± 4.9% vs. 25.0 ± 2.3%, $p < 0.001$). This finding was similar in averaged WT between HCM and control groups (11.3 ± 5.1 mm vs. 6.0 ± 2.3 mm, $p < 0.001$).

Table 1 Study population characteristics

	HCM(n = 125)	Control (n = 58)	p-value
Age (y)	58. ± 16	51. ± 14	p < 0.05
BSA (kg/m^2)	1.77 ± 0.2	1.67 ± 0.2	p = 0.026
ECV	27.5 ± 4.9	25 ± 2.3	p < 0.001
WT (mm)	11.3 ± 5.1	6 ± 2.2	p < 0.001
LV MASS	170.4 ± 76.2	97.3 ± 31.1	p < 0.001
LVEDV	94.7 ± 26.4	102.7 ± 22.6	p < 0.001
LVESV	18.5 ± 17	25.9 ± 12.5	p < 0.001
nLVEDV	53.6 ± 14.6	60.7 ± 11.9	p < 0.001
nLVESV	10.5. ± 10.1	15.3 ± 6.9	p < 0.001
CI	2876 ± 537	3313 ± 585	p < 0.001
LVEF	82.3. ± 9.8	75.9 ± 7.7	p < 0.001
LVPER	−4.9 ± 1.3	−3.8 ± 0.7	p < 0.001
LVPFR	3.9 ± 1.5	4.6 ± 1	p < 0.001

BSA, Body Surface Area; ECV, Extracellular Volume Fraction; WT, Myocardial Wall Thickness; LV, left ventricular; EDV, end-diastolic volume; ESV, end-systolic volume; nLVEDV, normalized LVEDV; CI, cardiac index; EF, ejection fraction; PER, peak ejection rate; PFR, peak filling rate.

B. Extracellular Volumes and Wall thickness

The regional ECV and regional WT measurements from the four different regions in each group are listed in Table 2. For regional comparisons, the septal region had the highest ECV compared with the other three regions in all the groups. To further investigate the correlation between the myocardial ECV and WT, we performed linear correlation analysis for each group. Our results demonstrated that the regional ECV significantly correlated with the regional WT in HCM group ($r = 0.381$, $p = 0.0001$) (Fig. 1). In contrast, no significant correlation ($r = 0.139$, $p = 0.297$) was found between regional ECV and WT in control group.

Table 2 The regional ECV and wall thickness in HCM group

	HCM(n = 125)	Controls(n = 58)	p-value
ECV(A) (%)	27 ± 4.9	24.4 ± 2.8	p < 0.001
ECV(S) (%)	28.5 ± 5.4	25.7 ± 2.3	p < 0.001
ECV(P) (%)	27.9 ± 5.6	25.3 ± 3.2	p < 0.001
ECV(L) (%)	26.5 ± 4.9	24.5 ± 2.6	p < 0.001
WT (A) (mm)	11.4 ± 6.7	4.8 ± 2.2	p < 0.001
WT (S) (mm)	14.5 ± 6.6	8 ± 3.1	p < 0.001
WT (P) (mm)	9.6 ± 5.0	5.5 ± 2.3	p < 0.001
WT (L) (mm)	9.5 ± 5.4	5.5 ± 2.2	p < 0.001

ECV, Extracellular Volume Fraction; WT, Myocardial Wall Thickness; A, anterior; S, septal; P, posterior; L, lateral.

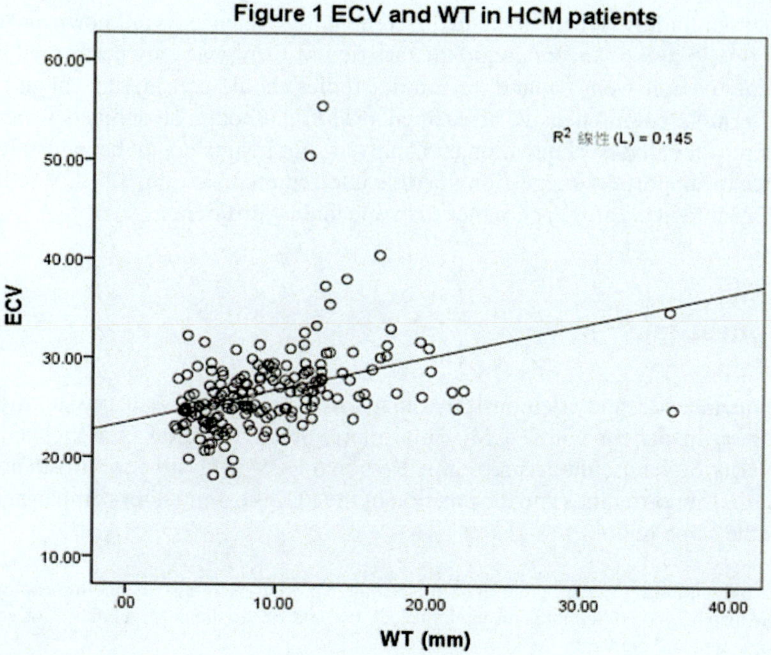

Fig. 1 Linear regression between the averaged ECV (extracellular volume fraction) and the averaged WT (wall thickness) in HCM group (N = 125)

4 Discussion

In this study, we demonstrated that CMR-derived ECV was significantly correlated with the myocardial wall thickness. This finding suggests that LVH is significant associated with the severity of DMF in patients with HCM. Myocardial fibrosis is defined as a significant increase in the collagen content in the myocardium. HCM are known to develop diffuse, as well as regional myocardial fibrosis. Once fibrosis develops, it increases myocardial stiffness and eventually deteriorates diastolic and systolic functions [17, 18].

Our results also showed that ECV varies with the myocardial region, suggested that regional ECV may lead to clinically significant errors if ECV was assessed and compared in different myocardial regions. Therefore, caution must be exercised in estimating the ECV in patients with significant ventricular hypertrophy.

5 Study Limitations

First, our study was done with a single T1 pulse sequence (MOLLI). Different sequences of T1 mapping have been reported to yield different absolute ECV values.

Whether our findings are identical in different pulse sequences is unknown and needs further investigation. Second, regional variation of ECV was only performed in one representative mid-ventricular slice, future studies should consider apical and basal slices for more comprehensive assessments. Third, although all subjects were carefully controlled in data acquisition and analysis, our results do not have histological evidence to support the suggestion that this interregional variation of ECV reflects a pathophysiological difference rather than a technical difference.

6 Conclusions

In conclusion, our study demonstrated that CMR-derived ECVs vary with myocardial region in patient with HCM. This interregional variation is associated with the severity of ventricular hypertrophy. Regional ECV is feasible for differentiating myocardial abnormalities and the position of the ROI is essential for comparing ROIs within the same region.

Acknowledgements I'm very grateful to my teacher, Mr. Mao-Yuan Su. He taught me a lot about MRI. He also helped me with this research project. I would like to thank Mr. Chiu for his constant encouragement and guidance.

Conflict of Interest The authors declare that they have no conflict of interest.

References

1. Maron BJ (2002) Hypertrophic cardiomyopathya systematic review. JAMA 287(10):1308–1320. https://doi.org/10.1001/jama.287.10.1308
2. Maron BJ et al (2003) American college of cardiology/European society of cardiology clinical expert consensus document on hypertrophic cardiomyopathy: a report of the American college of cardiology foundation task force on clinical expert consensus documents and the European S. J Am Coll Cardiol 42(9):1687–1713. https://doi.org/10.1016/S0735-1097(03)00941-0
3. Alcalai R, Seidman JG, Seidman CE (2008) Genetic basis of hypertrophic cardiomyopathy: From bench to the clinics. J Cardiovasc Electrophysiol 19(1):104–110. https://doi.org/10.1111/j.1540-8167.2007.00965.x
4. Seidman CE, Seidman JG (2011) Identifying sarcomere gene mutations in HCM: a personal history. Breast Dis 108(6):1–15. https://doi.org/10.1161/CIRCRESAHA.110.223834.Identifying
5. Richard P et al (2003) Hypertrophic cardiomyopathy distribution of disease genes, spectrum of mutations, and implications for a molecular diagnosis strategy. https://doi.org/10.1161/01.CIR.0000066323.15244.54
6. M. S.Maron *et al.*, "Hypertrophic Cardiomyopathy Phenotype Revisited After 50 Years With Cardiovascular Magnetic Resonance," *J. Am. Coll. Cardiol.*, vol. 54, no. 3, pp. 220–228, 2009, doi: https://doi.org/10.1016/j.jacc.2009.05.006.
7. Klues HG, Schiffers A, Maron BJ (1995) Phenotypic spectrum and patterns of left ventricular hypertrophy in hypertrophic cardiomyopathy: morphologic observations and significance as

assessed by two-dimensional echocardiography in 600 patients. J Am Coll Cardiol 26(7):1699–1708. https://doi.org/10.1016/0735-1097(95)00390-8
8. Maron BJ, Rowin EJ, Udelson JE, Maron MS (2018) Clinical spectrum and management of heart failure in hypertrophic cardiomyopathy. JACC: Heart Failure 6(5). Elsevier Inc., pp. 353–363. https://doi.org/10.1016/j.jchf.2017.09.011
9. Ellims AH, Iles LM, Ling LH, Hare JL, Kaye DM, Taylor AJ (2012) Diffuse myocardial fibrosis in hypertrophic cardiomyopathy can be identified by cardiovascular magnetic resonance, and is associated with left ventricular diastolic dysfunction. J Cardiovasc Magn Reson 14(1):1. https://doi.org/10.1186/1532-429X-14-76
10. Moon JC (2017) Left ventricular hypertrophy revisited cell and matrix expansion have disease-specific relationship. Circulation 136:2519–2521. https://doi.org/10.1161/CIRCULATIONAHA.117.029895
11. Lazzeroni D, Rimoldi O, Camici PG (2016) From left ventricular hypertrophy to dysfunction and failure focus reviews on heart failure. Circ J 80. https://doi.org/10.1253/circj.CJ-16-0062
12. Lovic D et al (2017) Left ventricular hypertrophy in athletes and hypertensive patients. J Clin Hypertens 19:413–417. https://doi.org/10.1111/jch.12977
13. Mewton N, Liu CY, Croisille P, Bluemke D, Lima JAC (2011) Assessment of myocardial fibrosis with cardiovascular magnetic resonance. J Am Coll Cardiol 57(8):891–903. https://doi.org/10.1016/j.jacc.2010.11.013
14. Kellman P, Hansen MS (2014) T1-mapping in the heart: accuracy and precision. https://doi.org/10.1186/1532-429X-16-2
15. Messroghli DR, Greiser A, Fröhlich M, Dietz R, Schulz-Menger J (2007) Optimization and validation of a fully-integrated pulse sequence for modified look-locker inversion-recovery (MOLLI) T1 mapping of the heart. J Magn Reson Imaging 26(4):1081–1086. https://doi.org/10.1002/jmri.21119
16. Wong TC et al (2012) Imaging association between extracellular matrix expansion quantified by cardiovascular magnetic resonance and short-term mortality. https://doi.org/10.1161/CIRCULATIONAHA.111.089409
17. RMA (1998) Pathologic fibrosis and connective tissue matrix in left ventricular hypertrophy due to chronic arterial hypertension in humans. J Hypertens Jul, 6(7):1031–1041. https://doi.org/10.1097/00004872-199816070-00018
18. Su MYM et al (2014) CMR-verified diffuse myocardial fibrosis is associated with diastolic dysfunction in HFpEF. JACC Card Imag 7(10):991–997. https://doi.org/10.1016/j.jcmg.2014.04.022
19. Brouwer WP et al (2014) In-vivo T1 cardiovascular magnetic resonance study of diffuse myocardial fibrosis in hypertrophic cardiomyopathy. J Cardiovasc Magn Reson 16(1):1–7. https://doi.org/10.1186/1532-429X-16-28

Upconversion Nanoformulation for Functional Imaging

Surojit Chattopadhyay, Sandip Ghosh, Najim Akhtar, De-Ming Yang, Chuan Lin Chen, and Ren-Shyan Liu

1 Introduction

Biomolecular imaging techniques is a powerful tool for diagnosis of cancers, and may have high specificity and sensitivity [1]. Among different varieties of metallic/metaloxide nanoparticles, upconversion nanoparticles (UCNPs) are superior for fluorescence imaging as they shows anti-Stokes shifted emission upon low energy NIR excitation, and are extremely stable even in physiological conditions [2]. The visible fluorescence that forms the basis of bioimaging with the UCNPs can be tuned from blue to red by changing the dopant type and concentration to suit multiplexing and has been reported before. The NIR excitation of these UCNPs result is negligible autofluorescence that increases the imaging contrast. The fluorescence in the UCNPs are also accompanied by non-radiative decay that results in heat in the UCNPs. This heat can also be imaged using a simple infrared thermal camera. On the other hand, the size and composition of the UCNPs promote easy cellular uptake, and offers minimum cytotoxicity. The easy incorporation of rare-earth ions such as Gd^{3+} and Yb^{3+} into UCNPs makes them active under magnetic resonance imaging (MRI) [3] and computed tomography (CT) [2]. Besides, they also provide an excellent platform for further functionalization [2] to combine with radio isotopes for single-photon emission computed tomography (SPECT) or positron

S. Chattopadhyay (✉) · S. Ghosh · N. Akhtar
Institute of Biophotonics, National Yang Ming Chiao Tung University, Taipei 112, Taiwan
e-mail: sur@nycu.edu.tw

D.-M. Yang
Microscopy Service Laboratory, Basic Research Division, Department of Medical Research, Taipei Veterans General Hospital, Taipei 112, Taiwan

C. L. Chen · R.-S. Liu
Department of Biomedical Imaging and Radiological Sciences, National Yang Ming Chiao Tung University, Taipei 112, Taiwan

K.-P. Lin et al. (eds.), *Future Trends and Challenges of Molecular Imaging and AI Innovation*, Springer Proceedings in Physics 272,
https://doi.org/10.1007/978-3-030-92786-8_5

emission tomography (PET) imaging, in order to improve the sensitivity of diagnosis. Such unique properties of these UCNPs makes them ideal candidate for the multimodal imaging which is essentially required to acquire biological information for identifying cancer at an early stage. UCNPs also found applications in photoenabled therapies. Overall, a control over the chemistry, surface functionalization, and knowledge of the photophysics in these UCNPs opens up a wide range of possibilities for biomedical applications.

In this work, we have done functional modification of the UCNPs with select proteins to achieve FRET which can be used for intracellular pH imaging. Next, we have modified the UCNP with another chelator (chl) protein and loaded radioisotopes therein to make the nanoformulation work as multimodal imaging agents.

2 Experimental Part

A. *Synthesis of UCNPs*

First, oleic acid (OA) capped UCNPs (dispersible in cyclohexane) were synthesized following a published report [4]. Typically, $GdCl_3$ (80%), $YbCl_3$ (18%), $ErCl_3$ (2%) were mixed with 14 mL of OA and 16 mL of octadecene in a three-necked flask, and heated to 150 °C under nitrogen atmosphere. Subsequently, 10 mL of methanol solution of NaOH (2.5 mmol) and NH_4F (4 mmol) is added dropwise while stirring for 30 min maintaining reaction temperature at 50 °C. Finally, the mixture was heated to 300 °C for 1 h, and washed with ethanol, and the precipitate redispersed in cyclohexane.

Next, the OA-capping was replaced with citrate (Cit-UCNPs) to make the UCNPs hydrophilic by ligand exchange reaction [5].

B. *Synthesis of mOrange fluorescent proteins and construction of UCNPs-mOrange FRET nanoprobes*

The His6-tagged mOrange fluorescent proteins were derived from pRSET-b mOrange plasmid. mOrange proteins were expressed in E. coli (BL21) competent cells. Later, we lysed the bacterial cell membrane, by performing mechanical lysis, using mechanical probe sonicator. Finally, we collected the proteins after centrifugation of the sample and purifird under known protocols. The Cit-UCNP-mOrange FRET nanoprobes were constructed via EDC (1-ethyl-3-(3-dimethylaminopropyl) carbadimide) (Sigma Aldrich, USA) chemistry [6]. For the FRET nanoprobe construction, we have used 2 mL of Cit-UCNPs (1.5 mg/mL) with 2 mL of mOrange FPs (0.35 mg/mL), and 3 mL of EDC (6 mg/mL) were incubated overnight at 4 °C with gentle shaking (200 rpm).

III. *Conjugation of radioisotope specific chelator (chl)*

1 mg of EDC was added into 5 mL aqueous solution of cit-UCNPs (0.2 mg/ml), and the solution was stirred at 340 rpm for 1 h at room temperature. 2 mg of as-purchased

Tc specific chelator was then added into the solution, and the solution was stirred under 200 rpm for 8 h at room temperature. The obtained UCNP@chl, formulation was collected by centrifuging the resultant mixture at 15,000 rpm for 30 min with the precipitation re-dispersed in 5 mL deionized water. $SnCl_2$ (1 mg/ml) and Tc-pertechnetate in DI water were sequentially added to the UCNP@chl solution and allowed to react for 30 min with shaking at room temperature. The precipitate of ^{99m}Tc labeled UCNP@chl (UCNP@chl-^{99m}Tc) was redispersed in 0.5 mL DI water for final use.

IV. *Characterization of UCNPs*

The morphology and crystal structures were imaged by High Resolution Transmission Electron Microscope (HR-TEM, JEM-2010F, JEOL, Japan). Fluorescence spectra were recorded by a Fluorolog spectrophotometer (JY Fluorolog 3 with iHR 320, Horiba, Japan), equipped with a 980 nm cw laser excitation (SDL, China). For the FRET imaging, we have used an inverted fluorescence microscope (Zeiss Axiovert 200) equipped with a 980 nm cw laser (CNI, China), and compatible optics and a wide-view dual-mode CMOS camera (Hamamatsu, Japan) to record high resolution images. 2 ml of as prepared UCNP@chl-^{99m}Tc (1 mg/mL) was taken in a quartz cuvette and 980 nm laser (SDL-5000 T, Shanghai, China) was used to observe the UCL. For the UCL imaging, the external 980 nm laser (SDL-5000 T, Shanghai China) along with optical fiber and a 10 mm laser collimator was installed in the IVIS (Xenogen IVIS 50, PerkinElmer, Inc. Massachusetts, U.S.A.).

E. *Cellular uptake and Intracellular pH imaging.*

For in situ calibrations of the pH nanoprobes, 100 μL of nanoprobe solution were incubated with HeLa cells (10^4 cells/well) for 24 h at 37 °C, and 5% CO_2. Later, intracellular pH was adjusted to that of the extracellular pH environment controlled by an ionophore 'Nigericin' (481,990, Calbiochem, Merck, USA). For this purpose, HeLa cells (incubated with UCNP-mOrange nanoprobes) were washed (by 1 × PBS) and incubated in pH (3.0, 5.0, and 7.0) controlled citrate/phosphate buffer medium. Afterwards, each well was treated with 20 μM of Nigericin and incubated for another 30 min at 37 °C, and 5% CO_2 before heading for the microscopic imaging.

3 Results and Discussion

A. Materials characterisations of UCNP, mOrange, and UCNP-mOrange FRET nanoprobes.

The UCNP formulation is schematically shown in Fig. 1. Figure 2 represents the morphological and FRET characteristics of UCNP-mOrange nanoprobes. Figure 2a shows the TEM image of Cit-UCNPs. The average size of this nanoparticles is 15 nm. Inset in Fig. 2a shows a colloidal solution of Cit-UCNPs (dispersed in DI water) emitting green upconversion luminescence (UCL) under 980 nm laser excitation.

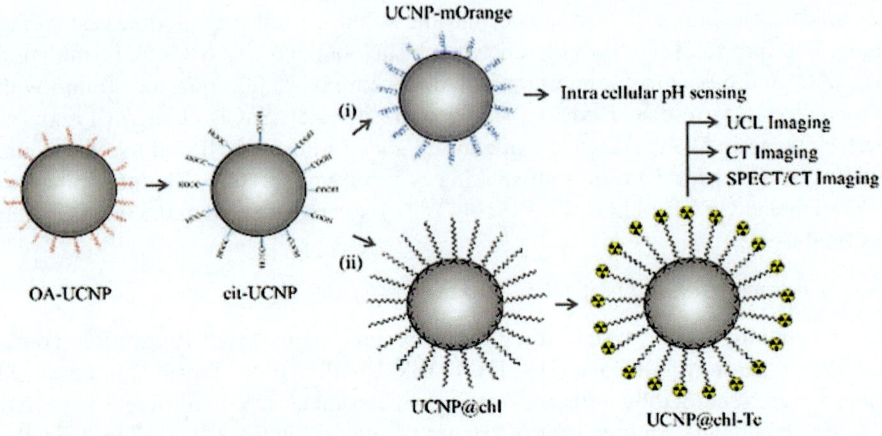

Fig. 1 Schematic of the UCNP formulation: (i) UCNP-mOrange for pH imaging, and (ii) UCNP-chl-Tc for multimodal imaging

Figure 2b, shows the HR-TEM image of a single UCNP-mOrange nanoprobe. Under TEM, the UCNPs (yellow circle) appears darker than a layer of mOrange protein coated outside (red circle). The overall size of the nanoprobe has increased to 25 nm because of the mOrange protein shell of 4.8 nm thickness around Cit-UCNPs.

Figure 2c shows the spectral overlap between UCNP (FRET donor) emission, and mOrange (FRET acceptor) excitation spectrum. We can observe that the broad excitation spectrum of mOrange fluorescent proteins is straddling the green emission band of Cit-UCNPs, satisfying the first condition of FRET. Figure 2d presents the spectroscopic FRET characteristics of UCNP-mOrange nanoprobe. Under the illumination of 980 nm laser, 5 distinct emission band has been observed. Out of these bands, peaks at 520, 521, 540, and 548 nm resembles to the characteristic peaks of UCNPs, and emission peak it 566 nm corresponds to the sensitized emission from mOrange fluorescent proteins. The confirmation of energy transfers between UCNPs and mOrange fluorescent protein is therefore established.

3.1 Intercellular pH Sensing Using UCNP-MOrange FRET Nanoprobe

Next, in addition to spectroscopic confirmation of FRET, we have investigated the in-vitro FRET imaging under different pH environment. To demonstrate the sensitivity of the nanoprobes to estimate the intracellular pH accurately, we have used Nigericin to modulate the intracellular pH with that of extracellular medium. Figure 3 presents the FRET images of HeLa cells in different pH (7.0, 5.0, and 3.0) buffers. The use of Nigericin ionophore would ensure an identical pH value inside and outside of the cells.

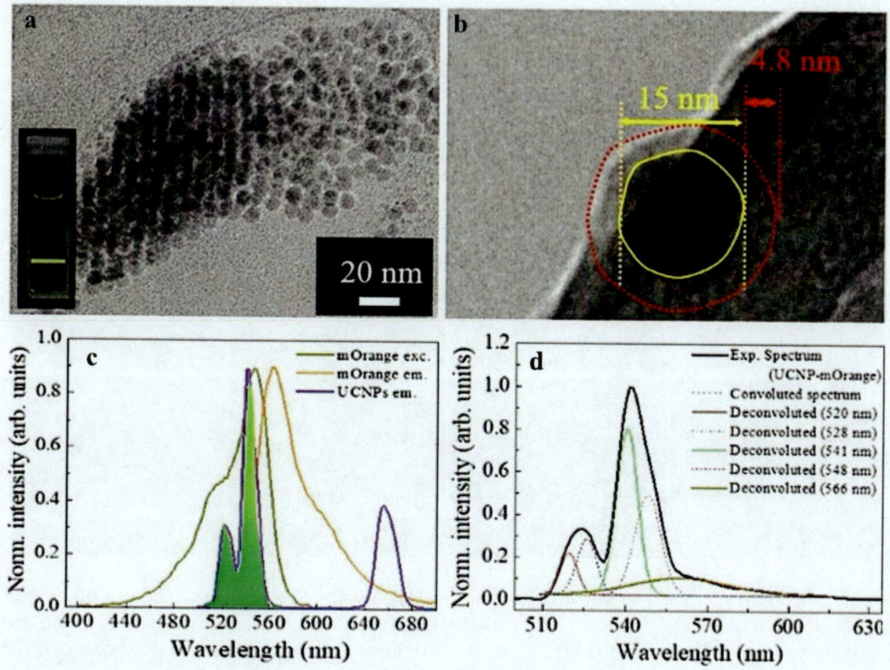

Fig. 2 Morphology and FRET characteristics of UCNP-mOrange nanoprobes. **a** TEM image of Cit-UCNPs. Inset shows a colloidal solution of Cit-UCNP (DI water) emitting green emission when illuminated with 980 nm laser. **b** High-resolution TEM image of a single UCNP-mOrange nanoprobe. The yellow circle shows the single Cit-UNCP, and a layer or mOrange protein with red circle. **c** Spectral overlapping between the emission of FRET donor (Cit-UCNPs), excitation of FRET acceptor (mOrange), respectively. **d** Convoluted and deconvolution FRET emission spectrum of UCNP-mOrange, under the excitation of 980 nm laser, which shows a sensitized emission band at 566 nm, and is coming from mOrange FPs. Reproduced with permission from Ref [7]. Copyright 2020 Elsevier B.V.

At alkaline pH (7.0, Fig. 3a), we could observe the individual signals from both UCNPs (green channel), and mOrange (red channel) from the nanoprobes uptaken in the HeLa cells. The FRET ratio (red to green channel signal) obtained is ~ 0.71. For the case of mildly acidic pH (5.0, Fig. 3b), the FRET intensity ratio dropped to ~ 0.52 as mOrange loses their ability to absorb energy from UCNP. At the high acidic pH (~3.0, Fig. 3c), the FRET ratio is observed to as low as ~ 0.30. Other than the drop in the FRET ratio at low pH, the overlay emission color has also been affected. The inactivity of the mOrange protein in acidic pH (3.0) would decrease the energy transfer efficiency making the emission green (Fig. 3c) dominated by the UCNPs. In basic pH, the emission will be relatively orange (Fig. 3a). The above results indicate that the UCNP-mOrange FRET nanoprobes can be used for accurate estimation of intracellular pH.

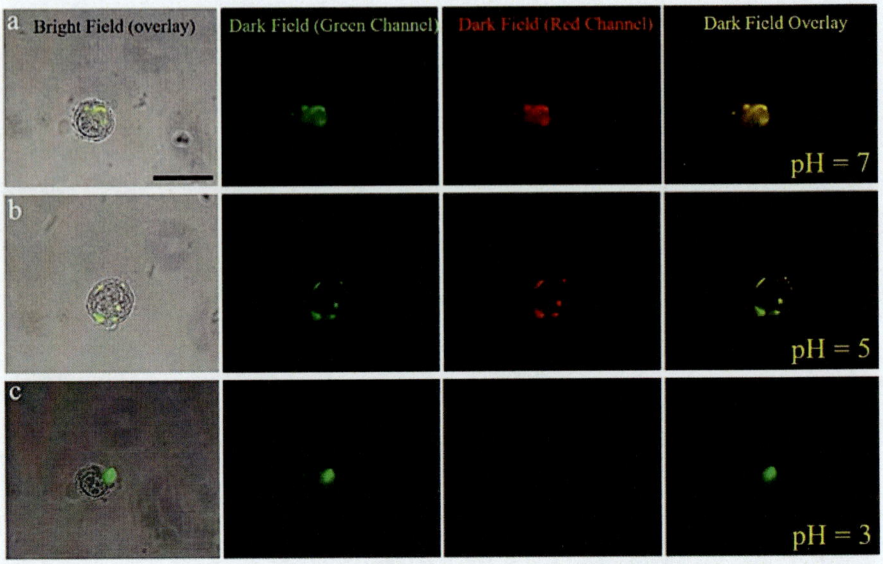

Fig. 3 Nigericin mediated in situ pH imaging in HeLa cells. Fluorescence imaging of UCNP-mOrange nanoprobes internalized HeLa cell(s) with nigericin modulated pH of **a** 7.0, **b** 5.0, and **c** 3.0, respectively. From left to right (L to R): Bright field image, green (UCNP) channel, red (mOrange) channel, and dark field overlay of the red and green channel images. Common scale bar of 10 μm. Reproduced with permission from Ref [7]. Copyright 2020 Elsevier B.V.

3.2 Multimodal Imaging Using UCNP-chl-^{99m}Tc Nanoformulation

Multi-modal imaging property of the UCNP@chl-^{99m}Tc is demonstrated by observing the UCL from the nanoformulation dispersed in water under 980 nm laser irradiation (Fig. 4a). UCL property of this nanocomposite was further investigated under small animal in vivo imaging system (IVIS) where only water containing glass vial shows no signal (Fig. 4b) under GFP filter, however, the vial containing 1 mg/mL of UCNP@chl-^{99m}Tc nanoformulation shows strong luminescence (Fig. 4c).

The Eppendorf containing the UCNP formulation without the radiolabel ^{99m}Tc (Fig. 4d) only shows the CT contrast because of the Yb^{3+} content in the formulation. The final formulation with the radio label (UCNP@chl-^{99m}Tc) shows strong SPECT signals (Fig. 4e). These results show proof-of-concept that the formulation could be used for multimodal imaging.

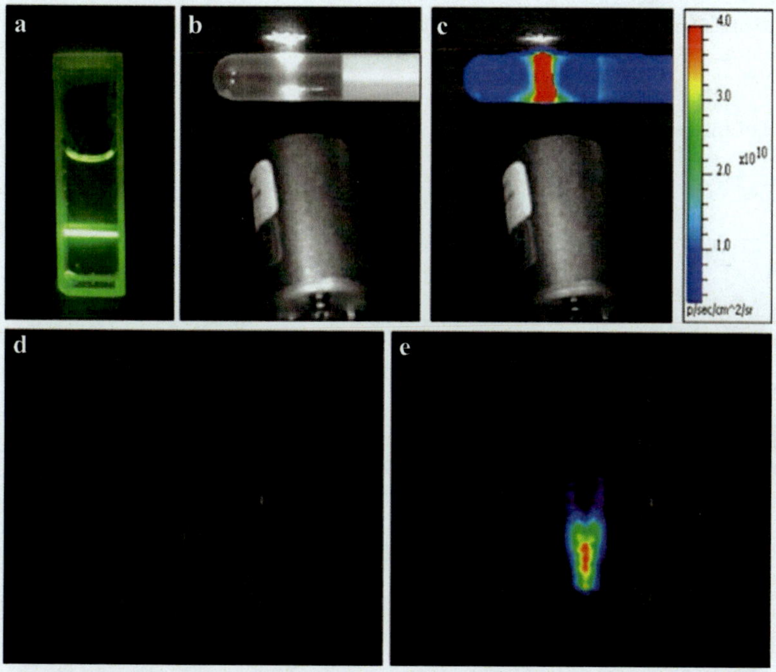

Fig. 4 Multimodal imaging of UCNP@chl-^{99m}Tc formulation. **a** Picture of quartz cuvette containing green fluorescent UCNP@chl-^{99m}Tc solution upon 980 nm irradiation. UCL image of glass vial containing **b** water, and **c** UCNP@chl-^{99m}Tc in water; laser power = 1 W. SPECT/CT image of **d** UCNP@chl and **e** UCNP@chl-^{99m}Tc

4 Conclusions

In conclusion, we demonstrate two UCNP based nanoformulation having functional imaging capabilities under deep penetrating 980 nm illumination. First, we have constructed a highly sensitive FRET based nanoprobe using UCNP as the donor and mOrange fluorescent protein as the acceptor. The UCNPs are internalized in HeLa cells. Upon changing the cellular pH by using an ionophore the emission from the UCNP-mOrange conjugate change from green at acidic pH and orange at alkaline pH. This UCNP based FRET imaging provides accurate estimation of intracellular pH. The results are reproducible, stable, and reversible indicating usability in early detection of cancer and other pH malfunctions in cells.

The second nanoformulation showed strong green fluorescence image contrast under 980 nm excitation. The component Yb^{3+} in the host $NaGdF_4$ of the UCNP makes the nanoformulation CT active. Radiolabelling of the UCNP obtained by chelating Tc in the protein network enabled SPECT/CT imaging. In short, the UCNP nanoformulation could demonstrate multimodal imaging.

Acknowledgements The authors would like to thank Ministry of Science and Technology, Taiwan (MOST-107-2112-M-010-003-MY3) for the research funding.

Conflict of Interest The authors declare no conflict of interest.

References

1. Ashok AH, Mizuno Y, Howes OD (2019) Tobacco smoking and dopaminergic function in humans: a meta-analysis of molecular imaging studies. Psychopharmacology 236(4):1119–1129
2. Gupta A, Lam CW, Wu CT, Yang DM, Chattopadhyay S (2018) Photothermal disintegration of 3T3 derived fat droplets by irradiated silica coated upconversion nanoparticles. Part Part Syst Charact 35:1800294
3. Bi H, He F, Dai Y, Xu J, Dong Y et al (2018) Quad-model imaging- guided high-efficiency phototherapy based on upconversion nanoparticles and ZnFe2O4 integrated graphene oxide. Inorg Chem 57(16):9988–9998
4. Ghosh S, Chiang WC, Fakhri MY, Wu CT, Chen RS, Chattopadhyay S (2020) Ultrasensitive broadband photodetector using electrostatically conjugated MoS_2-upconversion nanoparticle nanocomposite. Nano Energy 67:104258
5. Cao T, Yang T, Gao Y, Yang Y, Hu H, Li F (2010) Water-soluble $NaYF_4$:Yb/Er upconversion nanophosphors: synthesis, characteristics and application in bioimaging. Inorg Chem Commun 13:392–394
6. Dennis AM, Rhee WJ, Sotto D, Dublin SN, Bao G (2012) Quantum dot-fluorescent protein FRET probes for sensing intracellular pH. ACS Nano 6:2917–2924
7. Ghosh S, Chang YF, Yang DM, Chattopadhyay S (2020) Upconversion nanoparticle-mOrange protein FRET nanoprobes for self-ratiometric/ratiometric determination of intracellular pH, and single cell pH imaging. Bios Bioelectron 155:112115

Three-Dimensional Tomography of Cellulose Nanofibers- Polypeptides Nanocomposite Hydrogels

Tzu-Yi Yu, Yun-Hsiu Tseng, Ming-Chung Wu, Cheng-Si Tsao, and Wei-Fang Su

1 Introduction

In tissue engineering, researchers employ biomaterials to accelerate the rate of self-regeneration of damaged tissues [1–4]. This approach is practical if the biomaterials are similar to the environment supporting the target cells [5]. Extracellular matrix, ECM, is the matter between cells. It balances the chemical environment between cells, supports cell growth, and provides stimuli to guide cell differentiation. Most of the polymeric biomaterials for tissue engineering can hardly mimic both the microstructure and the biochemical environment of ECM. Recently, a type of nanomaterial derived from nature cellulose called "cellulose nanofiber" is an emerging ECM-mimic biomaterials. Cellulose nanofibers, CNFs, are synthesized by applying physical and chemical treatments to nature cellulose, such as woods and pulps [6]. This nanomaterial features its tunable crystallinity and surface chemistry according to the physical and chemical treatments during its synthesis [7–9]. Though many groups reported its excellent biocompatibility and applied CNFs hydrogels in tissue engineering, so far, these CNFs hydrogels crosslinked by cationic ions lack of biochemical stimuli leading to malfunction organoids [10–12]. Here, a new CNFs hydrogel system crosslinked with polypeptides is proposed to demonstrate the possibility of combining a synthetic polypeptide in CNFs hydrogels. In this research, the CNFs hydrogels were crosslinked with two types of crosslinkers—calcium cation (from calcium chloride, $CaCl_2$) and poly L-lysine-*random*-L-glutamic acid (PLLGA). The

T.-Y. Yu · Y.-H. Tseng · W.-F. Su (✉)
Department of Materials Science and Engineering, National Taiwan University, Taipei, Taiwan
e-mail: suwf@ntu.edu.tw

M.-C. Wu
Department of Chemical and Materials Engineering, Chang-Gung University, Taoyuan, Taiwan

C.-S. Tsao
Institute of Nuclear Energy Research, Longtan, Taoyuan, Taiwan

K.-P. Lin et al. (eds.), *Future Trends and Challenges of Molecular Imaging and AI Innovation*, Springer Proceedings in Physics 272,
https://doi.org/10.1007/978-3-030-92786-8_6

effect of crosslinker types on the CNFs hydrogels' microstructure in different scale was studied using four imaging techniques: small angle X-ray scattering (SAXS), scanning electron microscope (SEM), transmission X-ray microscope (TMX), and polarized optical microscope (POM).

2 Materials and Method

A. Nomenclature

CNF stands for "cellulose nanofiber", $PLL_{80}GA_{20}$ (or PLLGA) stands for the random copolymer of L-lysine and glutamic acid in the ratio of 4 to 1. CNF is the main component of hydrogels, and PLLGA is a synthetic polypeptide as one of crosslinkers to CNF hydrogel. The CNF hydrogel crosslinked by PLLGA is named CNF/PLLGA. Similarly, the CNF hydrogel crosslinked by $CaCl_2$ is named CNF/$CaCl_2$. POM is the abbreviation to "polarized optical microscope", and SAXS represents "small angle x-ray scattering".

B. Materials and Hydrogel Fabrication

The calcium chloride was used as purchased. PLLGA was synthesized by the ring-opening polymerization (ROP) and the hydrolysis of protecting groups. The detail of PLLGA synthesis refers to the literature [13]. The numbers of PLLGA repeating units were controlled around 100 ~ 150. The CNFs suspension was synthesized with TEMPO oxidation and $NaBH_4$ reduction. Before fabricating the CNFs hydrogels, the crosslinkers ($CaCl_2$ and PLLGA) were dissolved in DI water to prepare the crosslinker solution in 50 mM effective charge concentration.

The CNFs hydrogels were fabricated with the dropped method reported in the literature about CNFs hydrogels [10, 12, 14]. The crosslinker solution was gently dropped in the container with CNFs suspension to form the CNFs hydrogels. After immersion for one night, the supernatant was removed, and the CNFs hydrogel was immersed in DI water for another night to remove the excessive crosslinker. The CNFs composite hydrogels were prepared with 1 ml of CNFs suspension and 2 ml of crosslinker precursors.

C. Polarized Optical Microscope

The hydrogels were observed under POM (DM 2500 M, LEICA). To prevent the hydrogels from desiccation, they were immersed in DI water before characterization. In this research, the hydrogels were characterized under two modes: "Cross Polar" and "Cross Polar + Compensator". In cross-polar mode, the polarity of analyzer was vertical to of polarizer. While the black region represents amorphous or isotropic crystal phases, the white area represents anisotropic structure from oriented polymer chain or, in this study, nanofibers. By adding a compensator in the light path, the orientation of microstructures was distinguished by their colors. Purple regions were amorphous or isotropic crystals. In blue areas, polymer chains or nanofibers were oriented from right-up to left-down in images. For those in orange areas, their orientation is from left-up to right-down.

D. Small Angle X-Ray Scattering

The transmission SAXS data of CNFs hydrogels were obtained at 23A work station in National Synchrotron Radiation Research Center, NSRRC. The beam energy was 15 keV, the sample-to-detector distance was 3.875 m, and the q range was from 0.005 Å-1 to 0.35 Å-1. For good SAXS results, hydrogels' desiccation should be prevented, and the hydrogel samples should be characterized in minutes. The 2D SAXS pattern was further converted into 1D integral by software. The background transmission and scattering image were necessary for 1D integration. The SAXS 1D integral can be fitted by the following Eq. (1), which combines Ornstein–Zernike model and wormlike chain model [15–17].

$$I(q) = \frac{l_1}{1 + q^2\xi^2} + p(q) \tag{1}$$

E. Transmission X-Ray Microscope

The transmission X-ray microscope data of CNFs hydrogels were obtained at the 1B work station in National Synchrotron Radiation Research Center, NSRRC, Taiwan. For the X-ray tomography study, the aqueous hydrogel sample was stained with 0.1% w/v ruthenium tetroxide in ddH2O to enhance the hydrogels' contrast. After 3 h of staining, the sample was washed with methanol. Tear off a hydrogel piece with a thickness of about 14 ~ 17 μm to fix on the holder by polyamide tape. 20 μl gold particle solution (400 ~ 500 nm) was added to the hydrogel as the rotation target. The sample was rotated from -75° to 75°. The 3D structure was reconstructed by computer simulation through 151 cross-section images taken per degree by TXM.

3 Results and Discussion

A. Nanometer-scale Structure of CNF Hydrogels

SAXS is a nanostructure characterization method based on the x-ray scatter resulted from the electron density between phases in materials. Since CNF hydrogels comprise two phases, dense cellulose nanofibers and water in interconnected pores, the 2D scatter patterns present the reciprocal structure of CNF hydrogels network. In literature, a fibrous hydrogel network can be fitted with Eq. (1) [17]. The first term in equation represents the structure factor (fibers network structure), and the slope in low q region (q < 0.01 $Å^{-1}$) is related to the hydrogel mesh size (ξ). The smaller mesh size (ξ) results in the slope in low region (b) closer to -2. In Fig. 1a, the slope of CNF/PLLGA is close to the one of pristine CNF, high than -1. However, the CNF hydrogel crosslinked by $CaCl_2$ exhibits a slope close to -2. It indicates that $CaCl_2$ leads to a CNF network with a smaller mesh size (ξ). This structural change is a

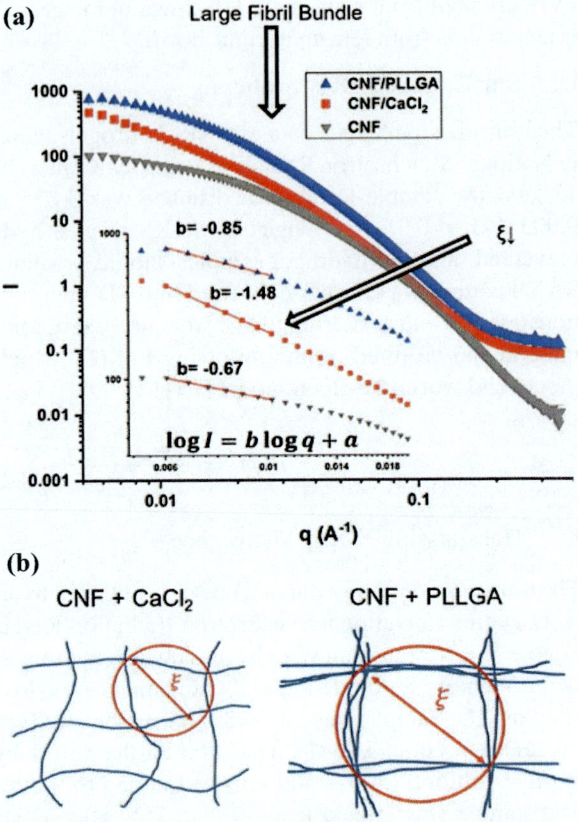

Fig. 1 Microstructure of CNF hydrogel in nanometer scale. **a** SAXS 1D integral of CNFs hydrogels crosslinked by different crosslinker precursors under 50 mM effective charge concentration. The slope at low-q region is related to the mesh size of network, ξ. **b** Simplified illustration of CNF/$CaCl_2$ and CNF/PLLGA. CNF assembles into bundles after adding PLLGA while CNF forms dense network after adding $CaCl_2$

typical result when polymer backbones are crosslinked by adding crosslinkers. On the other hand, the CNF/PLLGA SAXS result suggests a different hydrogel network structure from typical polymeric hydrogels. The second term in Eq. (1) is credited to the morphology of individual CNFs. Because it shows stronger scattering in middle q region, CNF/PLLGA SAXS result leads to a hypothesis that CNFs assembled into bundles when they crosslinked by PLLGA. Two CNF hydrogel networks resulted from either $CaCl_2$ or PLLGA are summarized in Fig. 1b. The network structure difference might lead to different physical properties, such as mechanical strength and water content.

B. Micrometer Scale Structure of CNF Hydrogels

Freeze-dried hydrogel samples are commonly used for structural analysis under scanning electron microscope. The morphologies of dried CNF/$CaCl_2$ and CNF/PLLGA hydrogels are shown in Fig. 2a and b, respectively. In SEM images, the type of crosslinkers has no significant effect on hydrogel structure in network morphologies. However, these images might not present the real network structure due to the distortion of hydrogel structure during sample preparation. To prepare SEM imaging

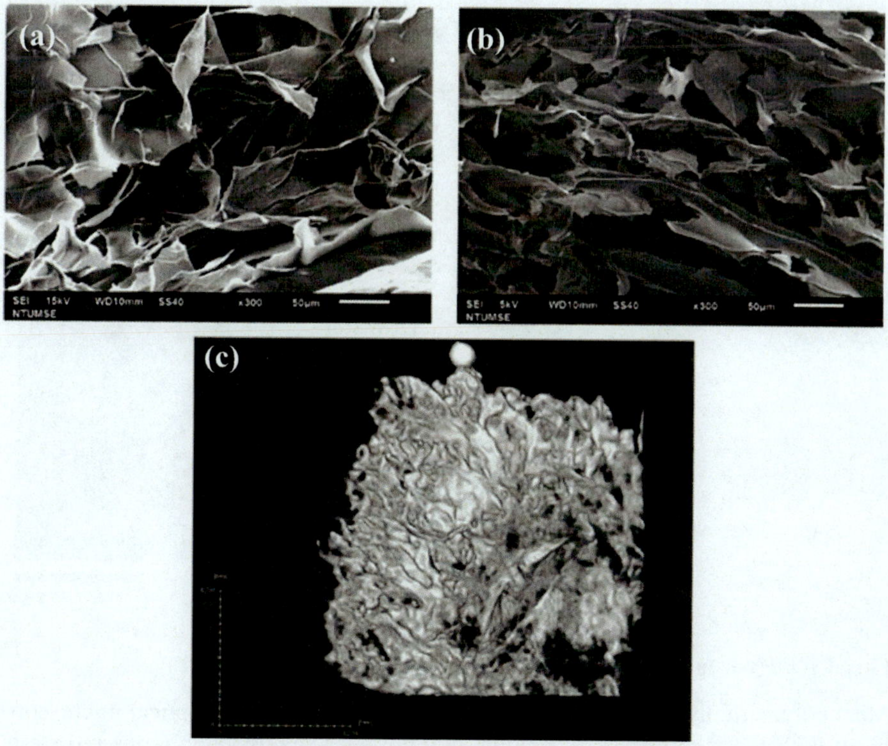

Fig. 2 Interconnected porous structure in CNF hydrogels in the micrometer scale. SEM image of (**a**) CNF/$CaCl_2$ hydrogels and (**b**) CNF/PLLGA. Scale bar represents 50 μm in (**a**) and (**b**). (**c**) Reconstructed 3D model of CNF/$CaCl_2$ hydrogel from Transmission X-ray Microscope. Scale bar represents 6 μm

sample, hydrogel samples have to be frozen and lyophilized to remove water. Ice crystallization brings distortion to its network structure. Scientist still use the pore size in hydrogel SEM image to estimate the hydrogel network density.

Transmission X-ray Microscope (TXM) is an emerging technique to construct hydrogel network images without distortion. In TXM, hydrogel samples are located on a rotatory holder, and the images are collected with different x-ray incident angles. 3D tomography structure of hydrogels can be constructed by compiling these images with a numerical method. In Fig. 2c, the 3D tomography model of CNF/$CaCl_2$ is shown. The hydrogel samples were immersed in a ruthenium tetroxide aqueous solution to enhance contrast. Thus, the region in white represents the interconnected pore filled with water. Comparing with morphology in Fig. 3a, CNF/$CaCl_2$ exhibits high network density and, unlike the flake structure in the SEM image, anomalous CNF domains in TXM. Because hydrogels can be observed by TXM under ambient conditions, the structural distortion during freeze-drying can be prevented. It leads to a realistic image of three-dimensional hydrogel networks in the micrometer scale.

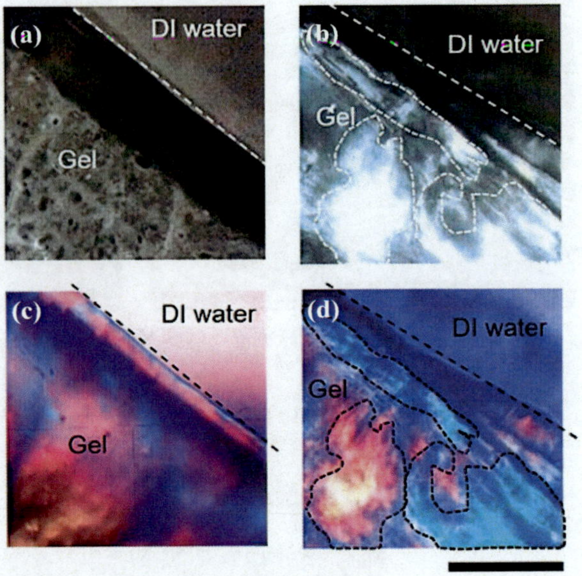

Fig. 3 Birefringence domains in CNF/PLLGA hydrogel induced by PLLGA. Cross-polar image of (**a**) CNF/$CaCl_2$ and (**b**) CNF/PLLGA hydrogel without the compensator. Cross-polar image of (**c**) CNF/$CaCl_2$ and (**d**) CNF/PLLGA hydrogel with compensator. Scale bar represents 250 μm

C. CNF Hydrogels under Polarized Optical Microscope

Most polymeric hydrogels have little optical structure under an optical microscope because they are composed of random coil polymer chains without ordered structures. However, some exhibit birefringence under POM for their order molecular orientation or liquid–crystal-like microstructure [18]. When CNFs are oriented into an ordered structure, CNF hydrogels show such property for the cellulose anisotropic crystal domain within CNF [7, 8, 19]. In Fig. 3b, there are several birefringence regions (white region) in CNF/PLLGA hydrogel, while DI water (background) shows no optical signal. By adding a compensator in the light path, these birefringence regions can be distinguished into two domains in either blue or orange, labeling two orientations. On the other hand, there is no significant birefringence observed in CNF/$CaCl_2$ in Fig. 3a and c. This result supports the previous hypothesis - CNFs assembles into bundles after crosslinked by PLLGA- because CNFs must form ordered structures for birefringence. In contrast, CNFs remain single nanofibers after adding $CaCl_2$, so it shows no birefringence in POM.

4 Conclusions

Four structure characterization methods, SAXS, SEM, TXM, and POM, were used to investigate the crosslinker effect on CNF hydrogel network in different scales. Based on SAXS results, we made a hypothesis that CNFs assemble into bundles after crosslinked by PLLGA. Its birefringence confirmed this hypothesis in POM images.

This microstructural difference might lead to a substantial difference in the physical properties of CNF hydrogels. We also demonstrated using TXM 3D tomography on CNF/$CaCl_2$ hydrogel to construct its in situ network structure without distortion during sample preparation.

Acknowledgements This research was supported financially by the Ministry of Science and Technology (MOST), Taiwan, under Contracts MOST 108–2221–E–002–027–MY3.

Conflict of Interest The authors declare that they have no conflict of interest.

References

1. Shekaran A, Garcia AJ (2011) Biochimica et Biophysica Acta (BBA)-General Subjects 1810(3):350–360
2. Lu H, Hoshiba T, Kawazoe N, Chen G (2011) Biomaterials 32(10):2489–2499
3. Pratt AB, Weber FE, Schmoekel HG, Müller R, Hubbell JA (2004) Biotechnol Bioeng 86(1):27–36
4. Sell S et al (2007) Polym Int 56(11):1349–1360
5. Xu C, Inai R, Kotaki M, Ramakrishna S (2004) Tissue Eng 10(7–8):1160–1168
6. Chirayil CJ, Mathew L, Thomas S (2014) Rev Adv Mater Sci 37(1–2):20–28
7. Shak KPY, Pang YL, Mah SK (2018) Beilstein J Nanotechnol 9:2479–2498
8. Isogai A, Saito T, Fukuzumi H (2011) Nanoscale 3(1):71–85
9. Khatri Z, Mayakrishnan G, Hirata Y, Wei K, Kim I-S (2013) Carbohyd Polym 91(1):434–443
10. Zander NE, Dong H, Steele J, Grant JT (2014) Acs Appl Mater Inter 6(21):18502–18510
11. Yang J, Zhang XM, Ma MG, Xu F (2015) Acs Macro Lett 4(8):829–833
12. Basu A, Lindh J, Alander E, Stromme M, Ferraz N (2017) Carbohyd Polym 174:299–308
13. Blout ER, Idelson M (1958) J Am Chem Soc 80(18):4909–4913
14. Masruchin N, Park B-D, Causin V, Um IC (2015) Cellulose 22(3):1993–2010
15. Geng L et al (2017) Cellulose 24(12):5417–5429
16. Mao Y, Liu K, Zhan C, Geng L, Chu B, Hsiao BS (2017) J Phys Chem B 121(6):1340–1351
17. Schoenmakers DC, Rowan AE, Kouwer PHJ (2018) Nat Commun 9(1):2172
18. Wu ZL et al (2011) Macromolecules 44(9):3542–3547
19. Parker RM et al (2016) ACS Nano 10(9):8443–8449

Deep CNN with Conditional Random Field in Breast Cancer Metastasis Detection

Meilan Jimmy Hasugian, Ping-Yu Wu, Teng-Bin Chang, Bang-Hung Yang, Ren-Shyan Liu, Mei-Fen Chen, and Kang-Ping Lin

1 Introduction

In 2020, WHO released the global health estimates and reported that breast cancer is one of the top 10 leading causes of death among female in most of countries [1]. Early cancer diagnosis plays significant role in improving patients' survival rate [2]. One essential step in early breast cancer diagnosis is to detect metastasis in lymph nodes (Fig. 1 [3]) through analyzing the hematoxylin and eosin (H&E) stained histopathology slides. Whole slide images (WSIs) have been widely used to identify the normal and tumor cells [4], even more in localizing the malignant lesions (Fig. 2) [5]. However, examining extra-large WSIs (e.g., 20 GPixels) would impede the diagnosis process.

In this decade, the utilization of deep convolutional neural network (deep CNN) in computer vision problem has increased rapidly with a broad range from natural images to medical images [5].

M. J. Hasugian · P.-Y. Wu · T.-B. Chang · M.-F. Chen · K.-P. Lin (✉)
Department of Electrical Engineering, Chung Yuan Christian University, Taoyuan, Taiwan
e-mail: kplin@cycu.edu.tw

B.-H. Yang · R.-S. Liu
Department of Biomedical Imaging and Radiological Sciences, National Yang Ming Chiao Tung University, Taipei, Taiwan

Department of Nuclear Medicine and National PET/Cyclotron Center, Taipei Veterans General Hospital, Taipei, Taiwan

R.-S. Liu
Department of Nuclear Medicine, Cheng Hsin General Hospital, Taipei, Taiwan

M.-F. Chen · K.-P. Lin
Technology Translation Center for Medical Device, Chung Yuan Christian University, Taoyuan, Taiwan

K.-P. Lin et al. (eds.), *Future Trends and Challenges of Molecular Imaging and AI Innovation*, Springer Proceedings in Physics 272,
https://doi.org/10.1007/978-3-030-92786-8_7

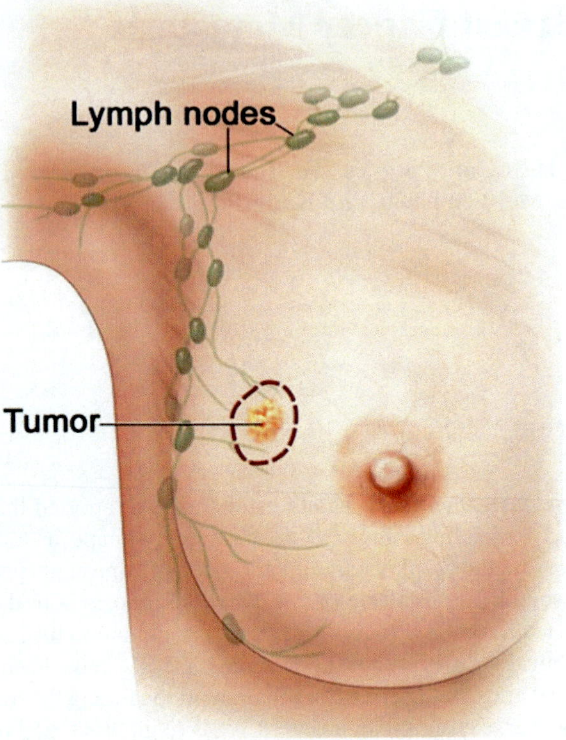

Fig. 1 Tumor and lymph nodes in breast cancer

Here, we present a deep learning-based approach combined with conditional random field for the identification of tumor metastasis from whole slide images of breast sentinel lymph nodes. In this study we compare performance several state-of-the-art CNNs models as feature extractor and combined with conditional random field (CRF) on the top as a framework to detect tumor metastasis in WSIs. Due to the extremely large size of the WSIs, every slide is segregated into numerous patches and perform the classification process on each patch independently. CRF is employed to preserve the spatial correlations between the adjacent classified patches.

To obtain the goals, every patch is divided into several sub-patches to retain the spatial connectivity during the training process. Then from these patches level predictions, tumor probability heatmaps is obtained to make the slide-based predictions dan tumor localization task.

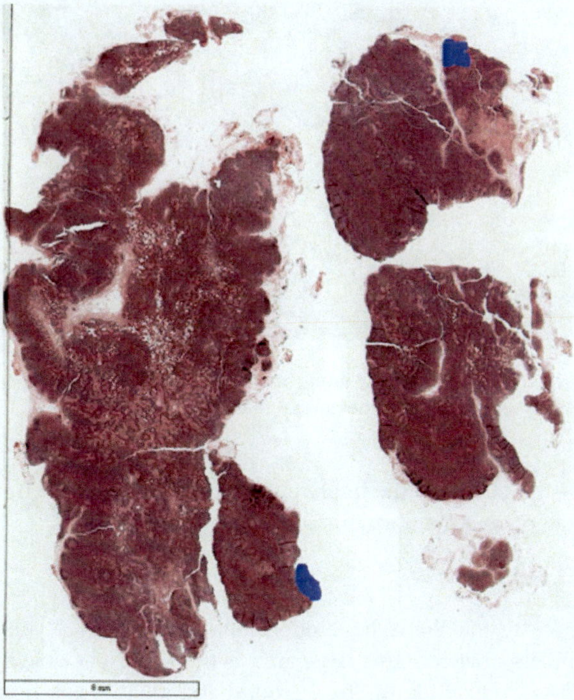

Fig. 2 Whole slide image with metastasis label (blue color)

2 Materials and Methods

A. Dataset

In this section, we describe the whole slide images provided in Camelyon16 dataset. The Camelyon16 dataset [6] consists of a total of 399 whole slide images (WSIs) split into 270 for training and 129 for testing. Both splits contain samples from two institutions (Radbound UMC and UMC Utrecht) with specific details provided in Table 1 [7].

The ground truth data for the training slides consists of a pathologist's delineation of metastatic tumor area on WSIs of sentinel lymph nodes. The ground truth (labels) data are provided in XML files format that contains annotated contours vertices of the metastases' location. Figure 3 shows the visualization of the tumor metastasis label on the top of WSI image.

Table 1 WSI in the Camelyon16 dataset

Institution	Train (normal)	Train (tumor)	Test
Radboud UMC	100 slides	70 slides	79 slides
UMC Utrecht	60 slides	40 slides	50 slides
Total	160 slides	110 slides	129 slides

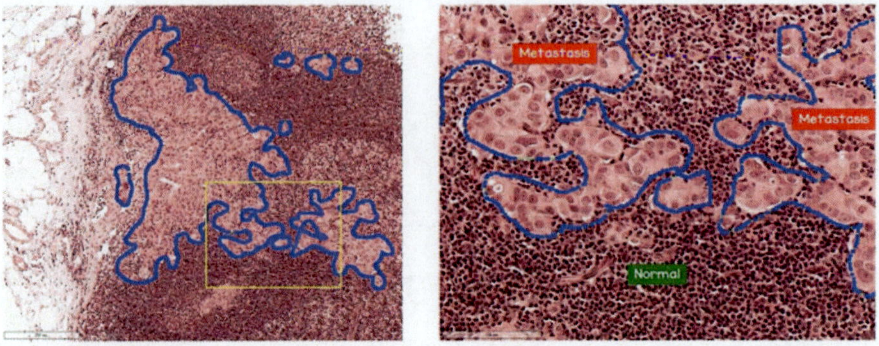

Fig. 3 Visualization of tumor metastasis label on the of a tumor slide image that contains the metastasis region and the normal region

As shown in the figure above, in the tumor slide image contains tumor metastasis region and the normal region.

B. Method

Mostly the WSIs have enormous numbers of pixels (around 100,000 × 200,000 pixels) that contains large area as the whitespace background. To focus the analysis, first we exclude the background and identify the tissue since the tumor metastasis exists in the tissue. This was achieved by implementing Otsu method [8] for each RGB channel then find the intersection between them.

Figure 4 shows the tissue area (right) from a slide image (left) as the focus in localizing the tumor metastasis. On average, approximately 85% of the WSI is the background region. Removing the background and aiming the process to the tissue area decreases the computation time.

To set up the patches for the training process, we randomly select 200,000 patches from tumor regions of tumor slides with size 768 × 768 pixels. For the normal patches we combined from normal slides and non-tumor regions from tumor slides total 200,000 patches with 768 × 768 pixels. The example of each patches is presented in Figs. 5 and 6.

Each patch is divided into nine sub-patches with size 256 × 256 pixels as the input of the model. In this study, we compare the impact of two frameworks in metastasis detection. The first framework is the baseline model with only using CNN as feature extractor then embedded the prediction into heatmap image as the prediction probability (Fig. 7).

In the feature extractor we deployed several well-known deep network architectures: VGG16 [9], GoogLeNet [10], ResNet-18 and ResNet-34 [11].

The second framework is depicted in Fig. 8. In this framework, a conditional random field (CRF) is added at the top of the baseline model. CRF is a probabilistic graphical model that used undirected graph representation and parameterization that encode a conditional distribution P(Y|X), where Y is a set of target variables given X

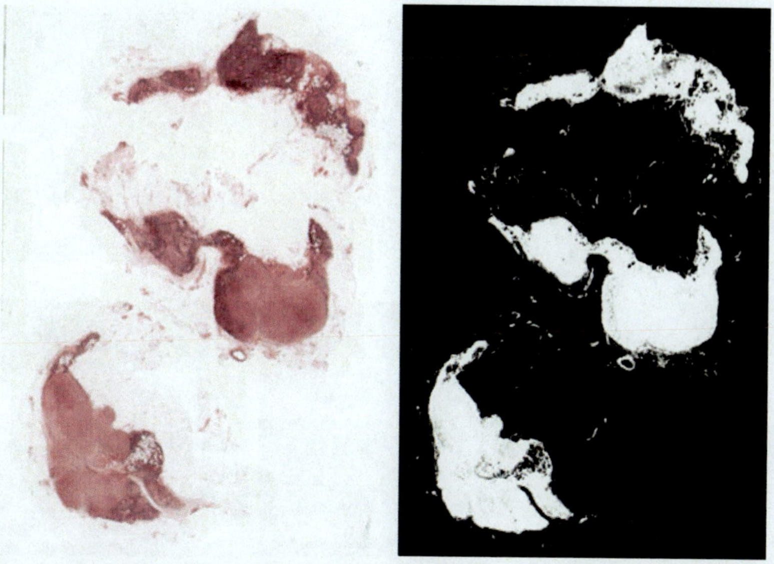

Fig. 4 The tissue area (right) as the focus of analysis, is obtained from the whole slide image (left)

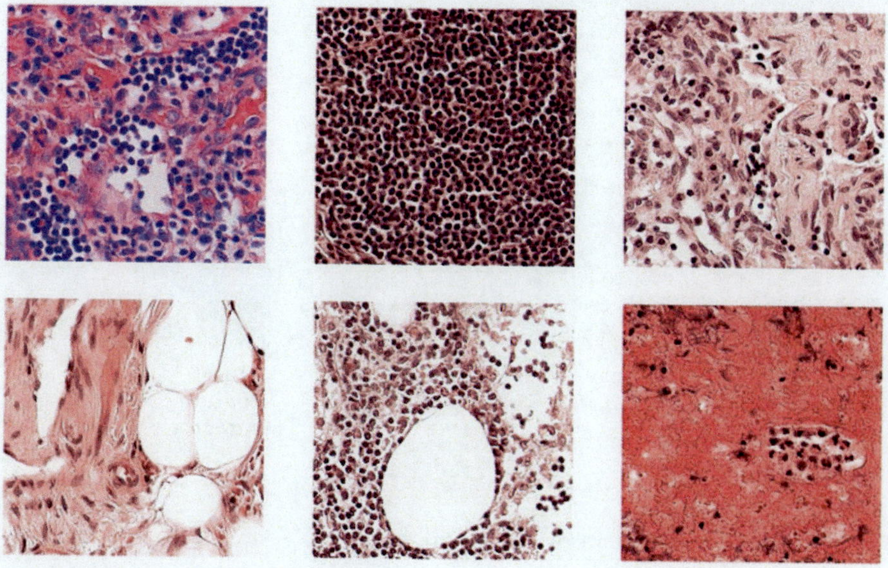

Fig. 5 Several patches from normal WSI

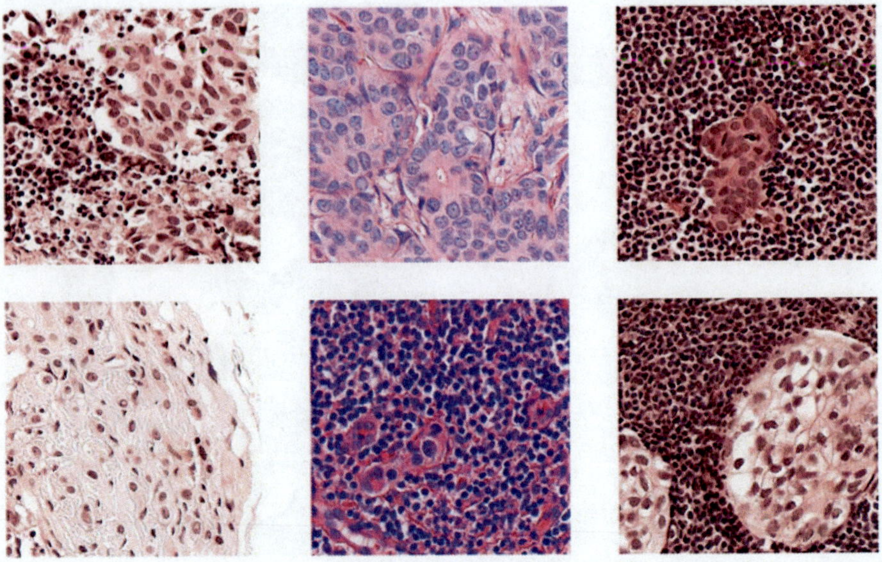

Fig. 6 Several patches from tumor WSI

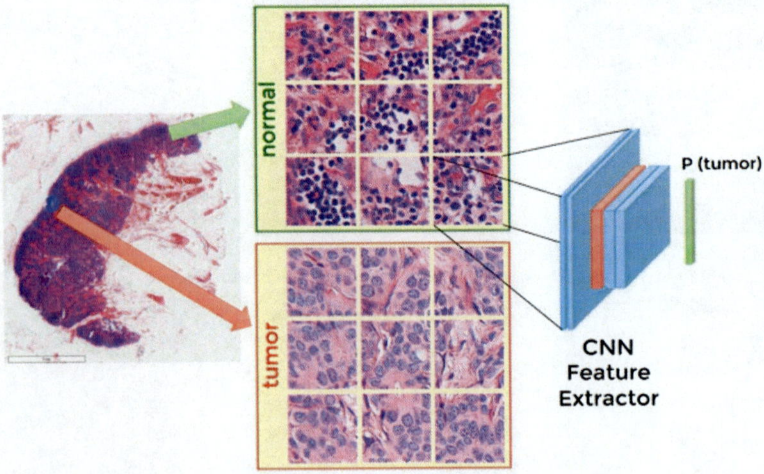

Fig. 7 The baseline model in metastasis detection

is a set of observed variables [12]. The image patches are divided into nine (N) sub-patches denoted as $x = \{x_i\}$ for $i = 1, \ldots, 9$. Let $Y = \{y_i\}$ be the random variables associated with sub-patches i, that represents the label of patch i takes a value of $\{normal, tumor\}$.

The conditional distribution P(Y|X) can be modeled as CRF with the Gibbs distribution of

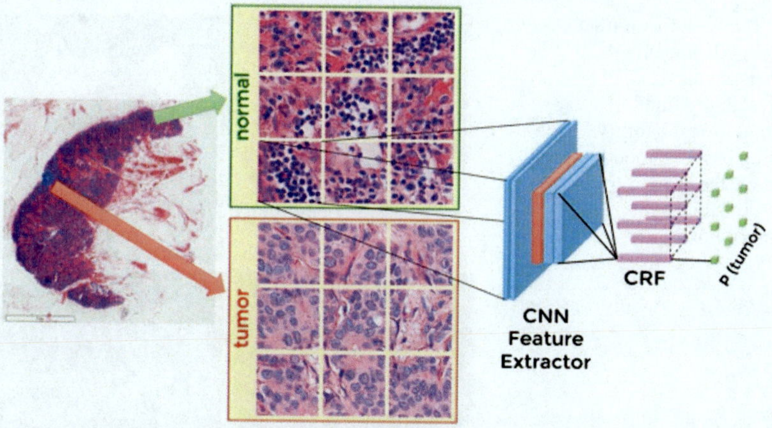

Fig. 8 The combination of deep CNN and CRF for metastasis detection

$$P(Y = y|x) = \frac{1}{Z(x)} e^{(-E(y,x))} \quad (1)$$

where E(y,x) is the energy function that measures the cost of Y taking a specific configuration y given x, and Z(x) is the partition function that insures P(Y = y | x) is a valid probability distribution. In the context of fully-connected CRF [13]. The energy function defined as:

$$E(y, x) = \sum_{i} \psi_u(y_i) + \sum_{i<j} \psi_p(y_i, y_j) \quad (2)$$

where i, j ranges from 1 to N. The unary potential $\psi_u(y_i)$ measures the cost of sub-patch i taking the label y_i given the sub-patch embedding x_i. The pairwise potential $\psi_p(y_i, y_j)$ measures the cost of jointly assigning sub-patch i, j with label y_i, y_j given the sub-patch embedding x_i, x_j. The pairwise potential models the spatial correlations between adjacent sub-patches.

3 Results

The framework was implemented with PyTorch 1.4.0 + cu92 and trained with NVIDIA GeForce RTX 2070 SUPER, 8 GB on the Ubuntu 20.04 LTS platform. The results of the baseline model with CRF on the slide Test_001 is displayed in Fig. 9. The patch classification accuracy of the baseline model and baseline + CRF on the validation slides dataset is presented in Table 2. The average free response receiver operating characteristic (FROC) score for the baseline model and baseline +

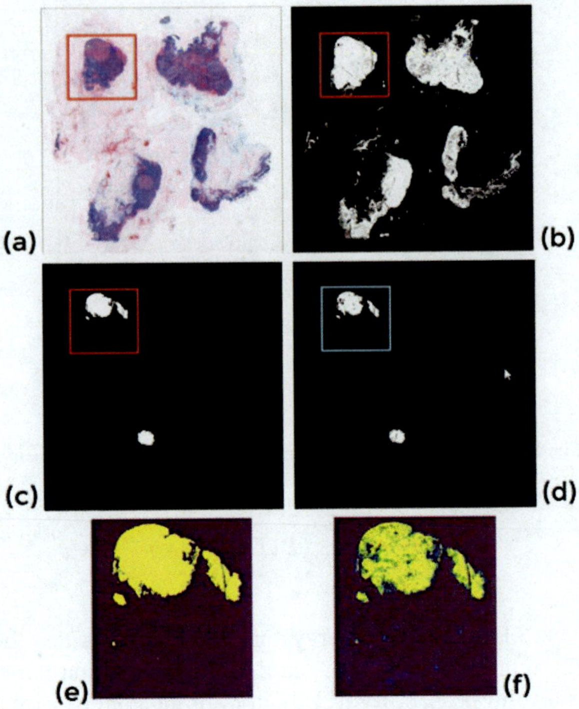

Fig. 9 The visualization on slide Test_001 **a** original slide, **b** tissue mask, **c** ground truth of tumor location, **d** the probability map using baseline model + CRF, **e** & **f** the ground truth and the probability map of the cropped box

Table 2 Patch classification accuracy on the validation slides

CNN	Baseline	Baseline + CRF
VGG16	0.8789 ± 0.0023	0.8809 ± 0.0021
GoogLeNet	0.8847 ± 0.0022	0.8924 ± 0.0021
ResNet-18	0.8903 ± 0.0022	0.8973 ± 0.0021
ResNet-34	0.8871 ± 0.0022	0.8910 ± 0.0021

CRF is reported in Table 3. The results show that the CNN baseline feature extractor is improved from the end-to-end joint training with CRF.

Table 3 Average FROC score on the test slides

CNN	Baseline	Baseline + CRF
VGG16	0.6989 ± 0.0113	0.6989 ± 0.0152
GoogLeNet	0.7102 ± 0.0112	0.7102 ± 0.0148
ResNet-18	0.7503 ± 0.0104	0.7503 ± 0.0139
ResNet-34	0.7391 ± 0.0107	0.7491 ± 0.0144

4 Conclusions and Discussion

In this study, we evaluated the influence of conditional random field on the top of CNN model for the detection of tumor metastasis on whole slide images. WSIs are the extra-large image with size around 20 GPixels. To reduce the computation time, each slide is randomly selected on the tumor and non-tumor region with size 768 × 768 pixels patch. This patch then divided into nine sub-patches with size 256 × 256 pixels to take the spatial correlation into consideration.

Compared to the baseline model, conditional random field enhances the ability of CNN feature extractor by integrating spatial correlations between adjacent sub-patches during the training process and improve the accuracy rate.

On the further development, increasing the numbers of patches image and including the borderline patch that contains tumor and non-tumor region is intuitively will enhance the accuracy together with taking more than nine sub-patches as well.

Conflict of Interest The authors declare that they have no conflict of interest.

References

1. WHO (2020) Global Health Estimates 2020: Deaths by Cause, Age, Sex, by Country and by Region, 2000–2019. Geneva, World Health Organization; 2020. https://www.who.int/data/gho/data/themes/mortality-and-global-health-estimates/ghe-leading-causes-of-death
2. Gecer B, Aksoy S, Mercan E et al (2018) Detection and classification of cancer in whole slide breast histopathology images using deep convolutional networks. Pattern Recogn 84:345–356
3. National Cancer Institute at https://www.cancer.gov/
4. Araujo T, Aresta G, Castro E et al (2017) Classification of breast cancer histology images using convolutional neural networks
5. Hou L, Samaras D, Kurc TM et al (2016) Patch-based convolutional neural network for whole slide image classification. In: IEEE conference on computer vision and pattern recognition 2016, pp 770–778
6. Camelyon Grand Challenge at https://camelyon16.grand-challenge.org
7. Bejnordi BE, Veta M et al (2017) Diagnostic assessment of deep learning algorithms for detection of lymph node metastases in women breast cancer. J Am Med Assoc:2199–2210
8. Otsu N (1979) A threshold selection method from gray-level histograms. IEEE Trans Syst Man Cybern 9(1):62–66
9. Simonyan K, Zisserman A (2015) Very deep convolutional networks for large-scale image recognition. In: ICLR 2015
10. Szegedy C, Liu W, Jia Y, Sermanet P et al (2015) Going deeper with convolutions. In: CVPR 2015
11. He K, Zhang X, Ren S, Sun J (2016) Deep residual learning for image recognition. In: Proceedings of the IEEE conference on computer vision and pattern recognition, pp 770–778
12. Koller D, Friedman N (2009) Probabilistic graphical models: Principles and techniques. MIT Press, Cambridge
13. Krähenbühl P, Koltun V (2011) Efficient inference in fully connected CRFS with Gaussian edge potentials. In: Advances in neural information processing systems, pp 109–117

Monitoring HDAC4 Expression in Alzheimer's Disease Using [^{18}F]TFAHA-PET

Yi-An Chen, Cheng-Hsiu Lu, Chien-Chih Ke, Chi-Wei Chang, Bang-Hung Yang, Juri G. Gelovani, and Ren-Shyan Liu

1 Introduction

Amyloid-β (Aβ) is widely considered as a key contributor to the pathophysiology of AD [1]. Despite various therapeutic approaches targeting Aβ through either secretase inhibitors or immunotherapy, the cognitive functions of patients failed to be

Y.-A. Chen · B.-H. Yang · R.-S. Liu
Department of Biomedical Imaging and Radiological Sciences, National Yang Ming Chiao Tung University, Taipei, Taiwan

Y.-A. Chen (✉)
Molecular and Genetic Imaging Core/Taiwan Mouse Clinic, National Comprehensive Mouse Phenotyping and Drug Testing Center, Taipei, Taiwan

C.-H. Lu
Core Laboratory for Phenomics and Diagnostics, Kaohsiung Chang Gung Memorial Hospital, Kaohsiung, Taiwan

Department of Medical Research, Kaohsiung Chang Gung Memorial Hospital, Kaohsiung, Taiwan

C.-C. Ke
Department of Medical Imaging and Radiological Sciences, Kaohsiung Medical University, Kaohsiung, Taiwan

Department of Medical Research, Kaohsiung Medical University Hospital, Kaohsiung, Taiwan

C.-W. Chang · B.-H. Yang · R.-S. Liu
PET center, Department of Nuclear Medicine, Taipei Veterans General Hospital, Taipei, Taiwan

J. G. Gelovani
Office of the Provost, United Arab Emirates University, P.O. BOX 1551, Al Ain, United Arab Emirates

R.-S. Liu
Department of Nuclear Medicine, Cheng Hsin General Hospital, Taipei, Taiwan

K.-P. Lin et al. (eds.), *Future Trends and Challenges of Molecular Imaging and AI Innovation*, Springer Proceedings in Physics 272,
https://doi.org/10.1007/978-3-030-92786-8_8

improved effectively [2]. Recent findings indicated the abnormal epigenetic modifications contribute to the impaired learning and deterioration of memory [3]. Epigenetic dyregulation causes the widespread decline in gene expression through post-translational histone modifications including epigenetic modifying enzymes, histone deacetylases (HDACs) [4]. Among these histone-modifying enzymes, HDAC4, belongs to HDAC class IIa enzymes, is highly enriched in brain and its homeostasis is associated with regulating the transcription of synaptic plasticity-related gene, neuronal survival and neuron development [5]. Studies have shown that the nuclear HDAC4 level is markedly increased in brains of AD patients and the total HDAC4 levels are highly expressed in AD mouse model [6, 7]. It suggests that abnormal HDAC4 expression or its nuclear localization may contribute to learning and memory deficits.

Given that the importance of HDAC class IIa in epigenetic regulation involved in brain function and HDAC inhibitor treatments restore cognitive performance, it is necessary to develop molecular imaging agents for evaluation of HDACs activity in the brain. Consequently, this has led to the development of a second generation of HDAC class IIa–specific radiotracer, 6-(tri-fluoroacetamido)-1-hexanoicanilide ([^{18}F]TFAHA), which shows significantly higher selectivity for HDAC class IIa enzymes [8]. Recent study has described that PET imaging with [^{18}F]TFAHA enable quantitative assessment of HDAC class IIa enzymes expression-activity in intracerebral 9L and U87-MG gliomas in rats [9]. However, whether PET imaging with [^{18}F]TFAHA can be used to monitor HDAC class IIa enzymes in AD brain needs to be investigated.

The aim of this study is to investigate the role of HDAC4 using 3D human neural cell model of AD by analysis of expression and treatment of specific inhibitor. We also evaluate HDAC4 expression in 3xTg AD mice compared with age-matched WT mice by performing PET imaging with [^{18}F]TFAHA and immunohistochemical analysis. The results suggest that [^{18}F]TFAHA is a useful tool for in vivo monitoring epigenetic deregulation, especially HDAC4, in AD progression.

2 Materials and Methods

A. Establishment of 3D human neural cell culture model of AD and relevant assays

3D culture models were set up for subsequently IF and biochemical analysis according to the method described previously [10]. To further investigate the effects of Tasquinimod on neuronal memory and synaptic plasticity-related genes, 3D-differentiated FAD cells were treated with Tasquinimod at doses between 30 and 100 μM for 48 h. Subsequently, RT-qPCR for gene expression was performed.

B. Animals

3xTg-AD mice (JAX-34830) and age-matched control group were used in this study. All animal procedures were performed in accordance with the institutional guidelines for care and use of laboratory animals.

C. Small animal PET/CT imaging experimental procedures

In PET image study, each AD mouse and age-matched WT mouse was injected intravenously with 8.04 ± 0.75 MBq/0.1 ml of [^{18}F]TFAHA. Subsequently, regional retention and uptake of these radioligands were processed and analyzed with PMOD 3.5 software package (Pmod Technologies, Zu¨rich, Switzerland).

D. Immunofluorescent staining

Brain Sects. (10 μm) or cell culture slides were incubated with the primary antibodies followed by fluorescence conjugated secondary antibodies (Abcam) and DAPI. All slides were observed under a confocal fluorescence microscopy (Zeiss LSM 880).

E. Statistical analysis

Data were statistically analyzed with GraphPad Prism (Version 5.0; GraphPad, La Jolla, USA) using Student's *t* test or by a one-way ANOVA, followed by Tukey's post hoc test.

3 Results

A. In vivo monitoring of HDAC4 using [^{18}F]TFAHA in 3xTg AD

To monitor HDAC4 expression in AD progression, we performed PET imaging with [^{18}F]TFAHA in AD transgenic mouse and age-matched wild type mouse. 3xTg AD mouse model, which exhibits progressive Aβ deposition and age-related changes in neuropathologies, is suitable to observe the changes of HDAC4 with age. Visual interpretation of [^{18}F]TFAHA-PET results showed greater uptake of [^{18}F]TFAHA in the whole brains of 3xTg AD mice compared with age-matched WT mice at 8, 11 and 16 months of age (Fig. 1a–c). Higher uptake were found in 3xTg AD mice consistently, suggesting that [^{18}F]TFAHA-PET imaging could be used to differentiate AD and WT control.

All brains were further anatomically parcellated according to a standard atlas and [^{18}F]TFAHA uptake in each region of cortex, hippocampus, striatum, basal forebrain, thalamus and cerebellum was evaluated as for whole brain analysis (Fig. 2). The 3xTg AD mice exhibited significantly higher [^{18}F]TFAHA uptake in most areas examined, including hippocampus ($t = 3.45$, $p = 0.04$), basal forebrain ($t = 5.61$, $p = 0.0112$), thalamus ($t = 4.36$, $p = 0.0224$) and cerebellum ($t = 2.38$, $p = 0.146$) compared to wild-type mice. Despite the [^{18}F]TFAHA uptake in neither cortex ($t = 2.67$, $p = 0.12$) nor striatum ($t = 4.06$, $p = 0.056$) reached statistical significance, both of these two regions of 3xTg AD mice showed higher [^{18}F]TFAHA uptake.

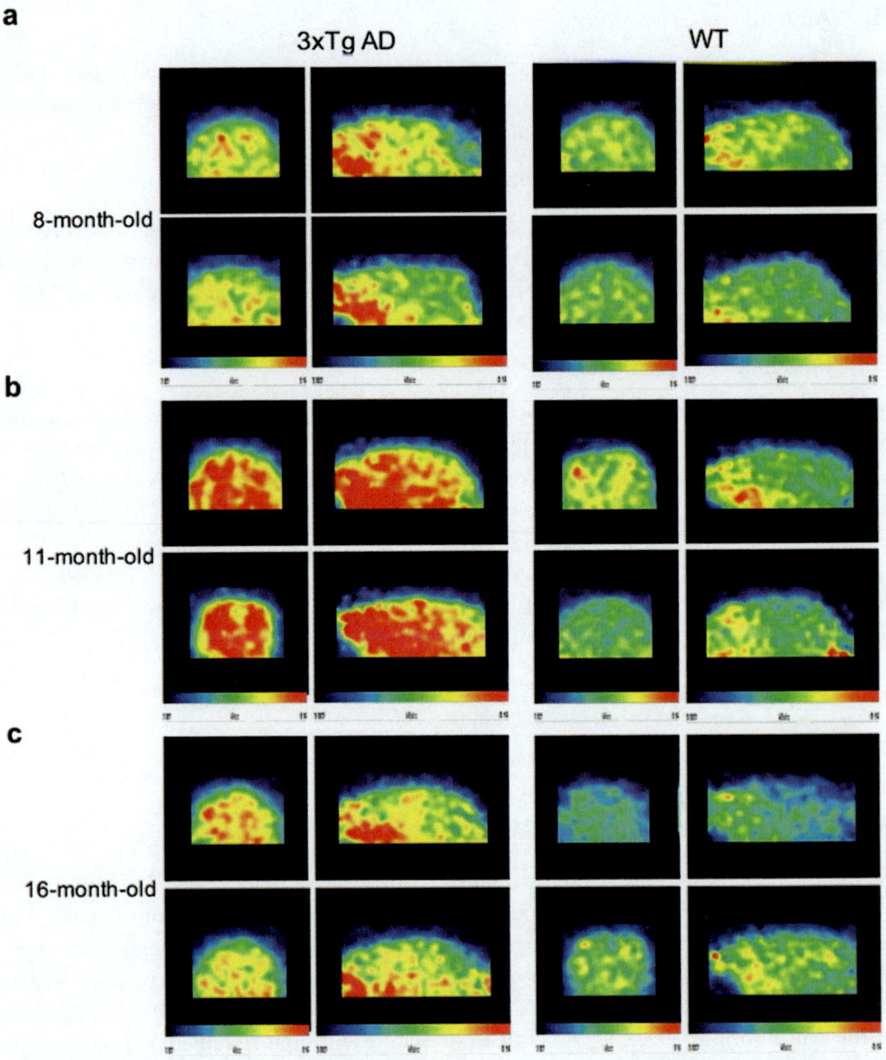

Fig. 1 In vivo PET imaging of HDAC4 using [^{18}F]TFAHA. Representative [^{18}F]TFAHA PET imaging of 3xTg AD mouse and WT mouse at **a** 8, **b** 11, **c** 16 months of age

To investigate whether the level of HDAC4 coincides with the uptake of [^{18}F]TFAHA, immunostaining with HDAC4 specific antibody was performed in brain sections. In line with the PET studies, differences in levels of HDAC4 were visually detectable between 3xTg AD and wild-type mice (Fig. 3).

B. In vitro characterization of HDAC4 expression in cell model of Alzheimer's disease

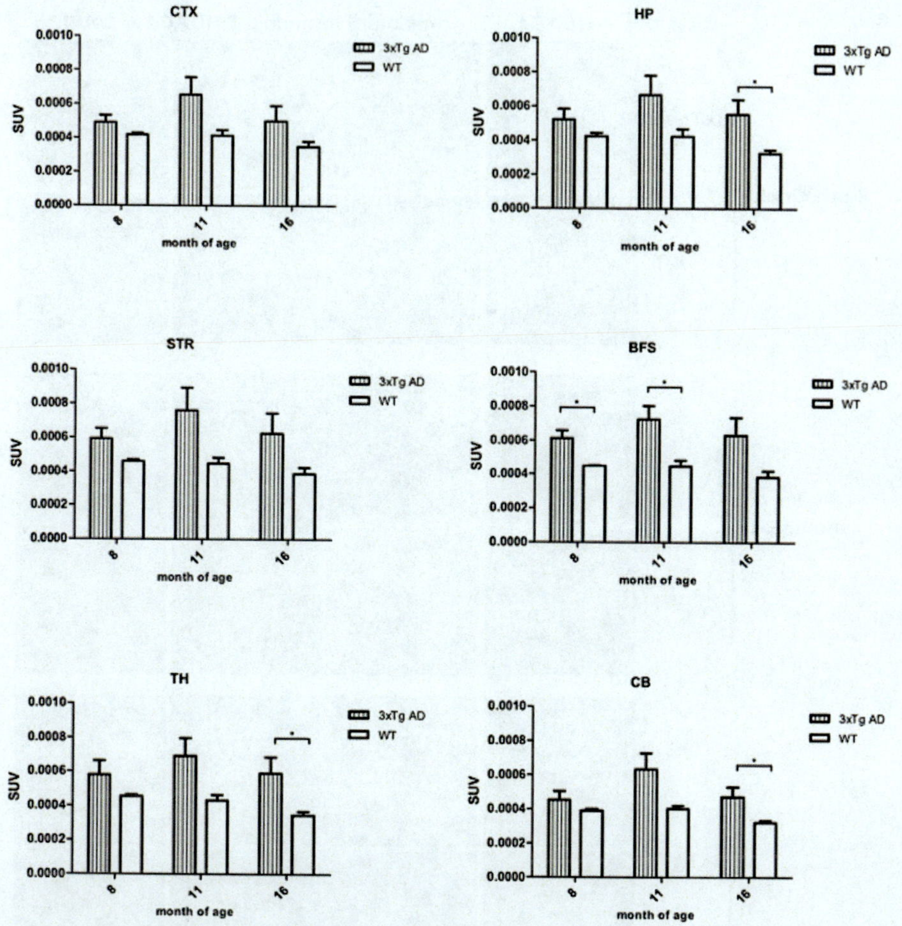

Fig. 2 The quantification of whole brain and different brain regions in SUV. SUV means standardized uptake value. Abbreviation: CTX, cortex; HP, hippocampus; STR, striatum; BFS, basal forebrain; TH, thalamus; CB, cerebellum

To investigate the HDAC4 expression following neuronal differentiation, a human neural cell culture model of AD that mimicked AD pathology by overexpressing human amyloid precursor protein with both Swedish (K670N/M671L) and London (V717I) FAD mutations in SH-SY5Y cell line was used (namely 3D-differentated FAD cells) in this experiment. IF staining and western blot analysis showed that HDAC4 was increased in 3D-differentated cells during neuronal differentiation either FAD or WT (Fig. 4a). By contrast, the level of HDAC1 remained largely unaffected. In addition, increased Hdac4 at the mRNA level was found in 3D-differentated FAD cells (Fig. 4b).

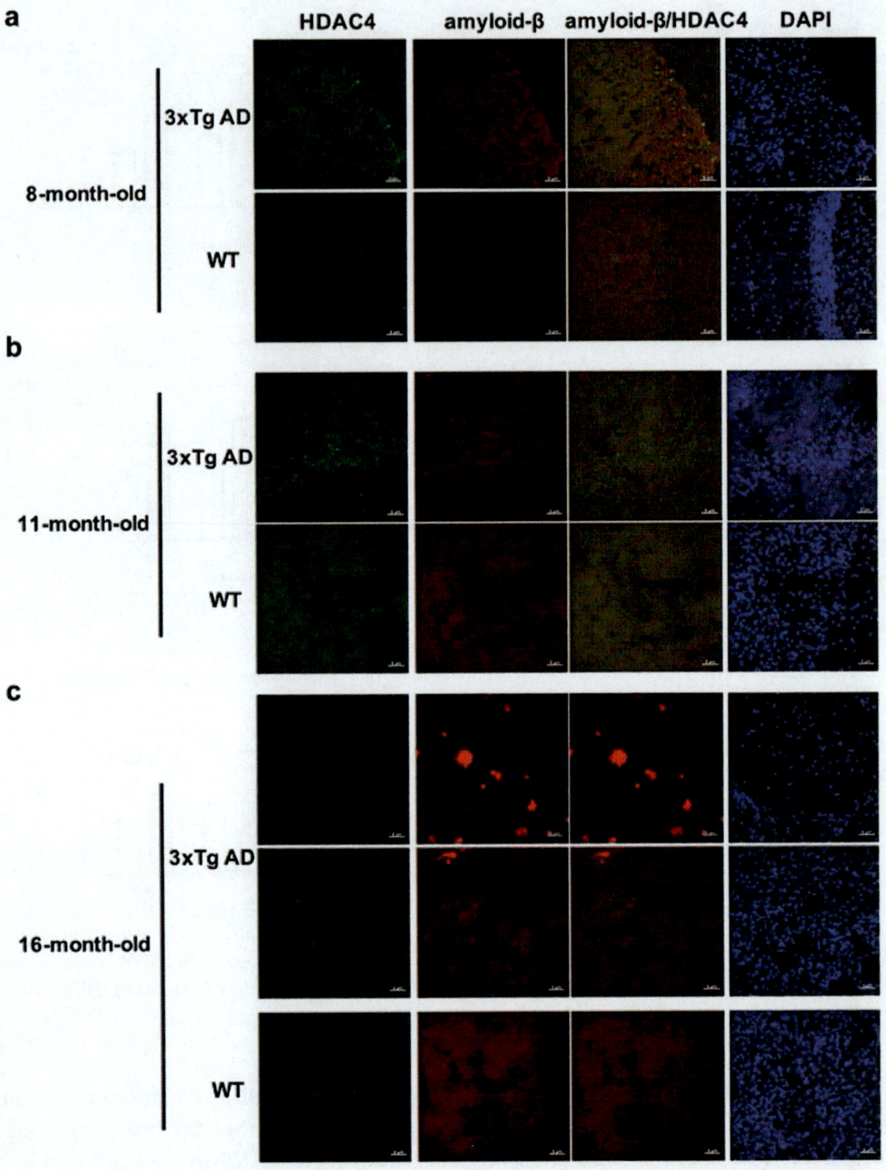

Fig. 3 Expression of HDAC4 and amyloid-β were increased in cortex of 3xTg AD mouse

C. HDAC4 inhibitor treatment up-regulates memory and synaptic plasticity-related genes

Given that the level of HDAC4 was elevated in cell model of AD, whether inhibition of HDAC4 can restore the levels of AD related affected genes is worthy of pursuit.

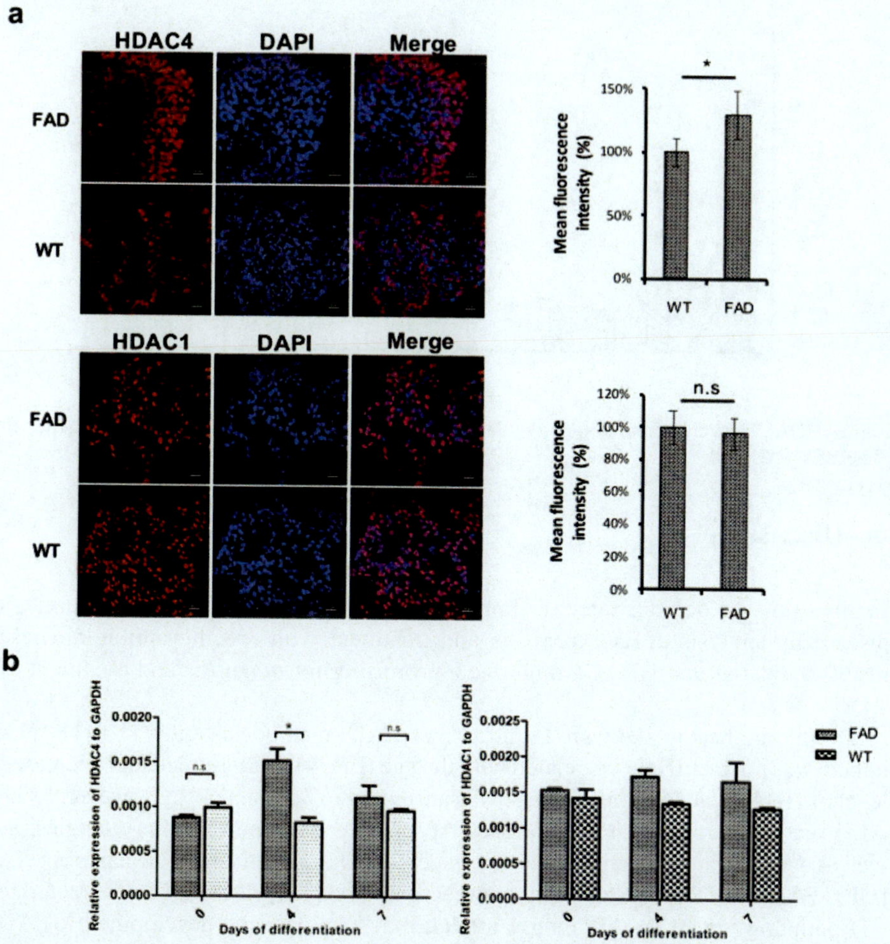

Fig. 4 The levels of HDAC4 are increased in 3D neural cell culture of AD

Tasquinimod (TasQ) is a highly specific inhibitor of HDAC4 and clinically tested oral antiangiogenic agent for therapy of castration-resistant prostate cancer. Herein, our study aimed to investigate the effect of TasQ on neuronal cells in AD. Quantitative RT-PCR analysis showed that TasQ upregulated several AD related affected genes and even dose-dependently increased the levels of Syn2, Homer1 and GluR2 (Fig. 5). Among these AD related affected genes, Lgi1, Syn2 and Homer1 belong to HDAC4 target genes [5]. Taken together, these data demonstrated that HDAC4 inhibitor treatment exhibited considerable therapeutic benefits on neuronal memory and synaptic plasticity-related genes in AD.

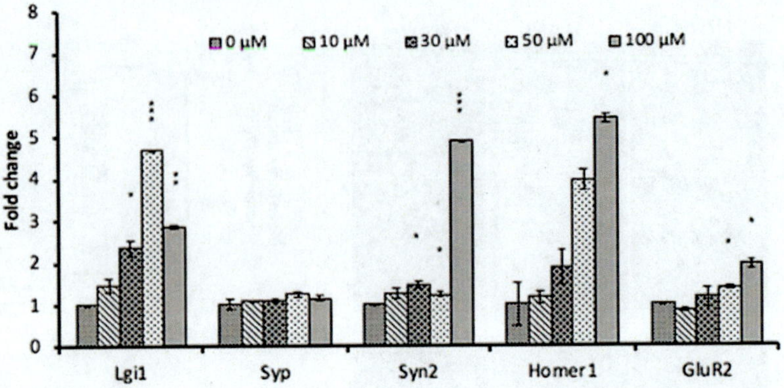

Fig. 5 HDAC4 specific inhibitor up-regulates memory and synaptic plasticity-related genes in SH-SY5Y FAD cells

4 Discussion

In this study, we demonstrated the important role that HDAC4 plays in Alzheimer's disease by analysis of its expression and treatment with specific inhibitor. These results validated that HDAC4 might be a promising for diagnosis and treatment of AD.

Despite higher uptake were found in 3xTg AD mice consistently, [^{18}F]TFAHA uptake was not positively correlated with the age (Fig. 4). We assumed that decreased level of HDAC4 at 16 months of age was attributed to neuronal loss. However, 3xTg AD mice were confirmed to expresses Aβ plaques and neurofibrillary tangles, as well as synaptic and behavioral deficits, except display significant neuronal loss [11]. [^{18}F]TFAHA-PET plus NeuN immunoreactivity and [^{18}F]FDG-PET to characterize AD pathology in 3xTg AD mouse model, may facilitate the development of AD research.

The variability of brain [^{18}F]TFAHA uptake at different age groups of AD mice was confirmed in current study. However, the progressive change of HDAC4 expression occurs in the same subject remains unknown. Thus, it is worthwhile to conduct longitudinal studies in mouse models of AD to better monitor and understand the events of epigenetic dysregulation throughout the course of the disease. Our data have shown that the level of HDAC4 coincides with the uptake of [^{18}F]TFAHA globally. However, owing to the limitation of μPET imaging for mouse brain, autoradiography assays can be used to evaluate the regional distribution and density of [^{18}F]TFAHA binding sites in the brains of the same mice that were imaged using PET.

There is growing evidence that aberrant HDAC expression and function leads to neuropathology. Previous studies showed that HDAC inhibitor specifically targeting class I, but not class IIa or IIb HDACs can significantly improve learning, memory and synaptic plasticity. Surprisingly, our results provided the first evidence that HDAC4 specific inhibition is also beneficial to AD related gene deficits. The therapeutic effect of HDAC4 inhibition in AD transgenic mice needs to be investigated in terms of behavior and neuronal function.

5 Conclusions

Current studies demonstrated that brain [^{18}F]TFAHA uptake at different age groups of AD mice coincided with the level of HDAC4. Moreover, a significant therapeutic effect of HDAC4 specific inhibitor on neuronal cells was demonstrated by upregulation of memory and synaptic plasticity-related genes. Furthermore, PET imaging with novel histone deacetylase class IIa selective substrate-based radiotracer, [^{18}F]TFAHA revealed HDAC4 expression in aging AD brains, which may encourage the development of AD-related neuroimaging and therapy.

Acknowledgements The authors thank Prof. Yi-Chao Lee (Taipei Medical University, Taiwan) for kindly providing the mouse line. We thank the Taiwan Mouse Consortium-Molecular and Genetic Imaging Center, which is funded by the Ministry of Science and Technology (MOST) of Taiwan for technical support in PET imaging experiment.

Conflict of Interest The authors declare that they have no conflict of interest.

References

1. Korczyn AD (2008) The amyloid cascade hypothesis. Alzheimers Dement 4(3):176–178. https://doi.org/10.1016/j.jalz.2007.11.008
2. Cummings J, Lee G, Mortsdorf T, Ritter A, Zhong K (2017) Alzheimer's disease drug development pipeline: 2017. Alzheimers Dement (N Y) 3(3):367–384. https://doi.org/10.1016/j.trci.2017.05.002
3. Penney J, Tsai LH (2014) Histone deacetylases in memory and cognition. Sci Signal 7 (355):re12. doi:https://doi.org/10.1126/scisignal.aaa0069
4. Esposito M, Sherr GL (2019) Epigenetic modifications in Alzheimer's neuropathology and therapeutics. Front Neurosci 13:476. https://doi.org/10.3389/fnins.2019.00476
5. Sando R 3rd, Gounko N, Pieraut S, Liao L, Yates J 3rd, Maximov A (2012) HDAC4 governs a transcriptional program essential for synaptic plasticity and memory. Cell 151(4):821–834. https://doi.org/10.1016/j.cell.2012.09.037
6. Shen X, Chen J, Li J, Kofler J, Herrup K (2016) Neurons in vulnerable regions of the Alzheimer's disease brain display reduced ATM signaling. eNeuro 3(1). https://doi.org/10.1523/ENEURO.0124-15.2016
7. Anderson KW, Chen J, Wang M, Mast N, Pikuleva IA, Turko IV (2015) Quantification of histone deacetylase isoforms in human frontal cortex, human retina, and mouse brain. PLoS ONE 10(5):e0126592. https://doi.org/10.1371/journal.pone.0126592
8. Bonomi R, Mukhopadhyay U, Shavrin A, Yeh HH, Majhi A, Dewage SW, Najjar A, Lu X, Cisneros GA, Tong WP, Alauddin MM, Liu RS, Mangner TJ, Turkman N, Gelovani JG (2015) Novel histone deacetylase class IIA selective substrate radiotracers for PET imaging of epigenetic regulation in the brain. PLoS ONE 10(8):e0133512. https://doi.org/10.1371/journal.pone.0133512

9. Laws MT, Bonomi RE, Kamal S, Gelovani DJ, Llaniguez J, Potukutchi S, Lu X, Mangner T, Gelovani JG (2019) Molecular imaging HDACs class IIa expression-activity and pharmacologic inhibition in intracerebral glioma models in rats using PET/CT/(MRI) with [(18)F]TFAHA. Sci Rep 9(1):3595. https://doi.org/10.1038/s41598-019-40054-2
10. Kim YH, Choi SH, D'Avanzo C, Hebisch M, Sliwinski C, Bylykbashi E, Washicosky KJ, Klee JB, Brustle O, Tanzi RE, Kim DY (2015) A 3D human neural cell culture system for modeling Alzheimer's disease. Nat Protoc 10(7):985–1006. https://doi.org/10.1038/nprot.2015.065
11. Virgili J, Lebbadi M, Tremblay C, St-Amour I, Pierrisnard C, Faucher-Genest A, Emond V, Julien C, Calon F (2018) Characterization of a 3xTg-AD mouse model of Alzheimer's disease with the senescence accelerated mouse prone 8 (SAMP8) background. Synapse 72(4). https://doi.org/10.1002/syn.22025

Study of a Position Decoding Method Making Imaging Area Continuous for a Planar Positron Imaging Probe

Hsin-Chin Liang, Chen-Han Cheng, and Hsiang-Ning Wu

1 Introduction

Due to population aging became serious in the recent decades, early diagnosis of neurodegenerative diseases have been noticed its clinical need. In such need, several PET tracers and their clinical trials have been showing promising results and early diagnosis potentials. Therefore, a project of developing a planar PET scanner for brain function imaging was raised. In our current design, the planar imaging probe is equipped with an imaging area of 151 $\times$ 151 mm^2, which obviously cannot cover a whole brain. Therefore, a sliding scan scheme is arranged for the brain scanner, as shown in Fig. 1. Due to the need of imaging probe sliding, it is required the whole imaging area of the proposed imaging probe to be successive. However, from the previous test result, as shown in Fig. 2, a 2D map of the imaging probe, lack of 2 rows/columns of crystals in the cross-over regions was observed. It is because of the conventional Anger logic not able to deal with the information from neighboring detectors [1]. Therefore, there is an urgent necessity for our group to build a position decoding algorithm capable of solving crossover problems between detectors [2], to make our imaging probe have a whole successive imaging area.

2 Materials and Methods

The proposed planar PET imaging probe is composed of 9 imaging detectors, connecting with a plate of 2.5-mm-thick window laying between the scintillation and photonics layer, shown in Fig. 3. Each detector is composed of a scintillation

H.-C. Liang (✉) · C.-H. Cheng · H.-N. Wu
Institute of Nuclear Energy Research/Health Physics Division, Taoyuan, Taiwan
e-mail: sjingliang@iner.gov.tw

K.-P. Lin et al. (eds.), *Future Trends and Challenges of Molecular Imaging and AI Innovation*, Springer Proceedings in Physics 272,
https://doi.org/10.1007/978-3-030-92786-8_9

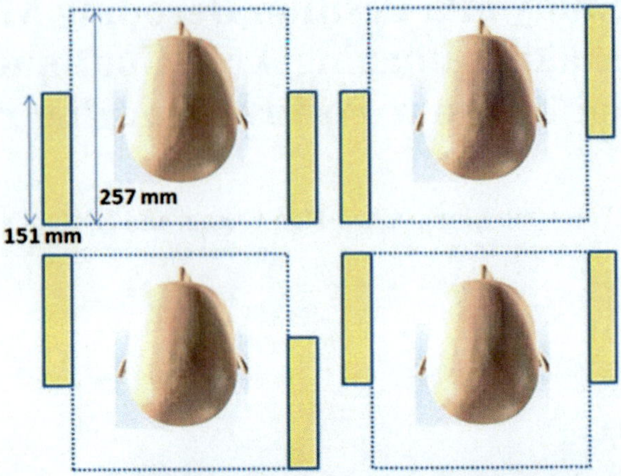

Fig.1 Sliding scan strategy for brain PET imaging

Fig.2 Crystal 2D map obtained with using Anger logic

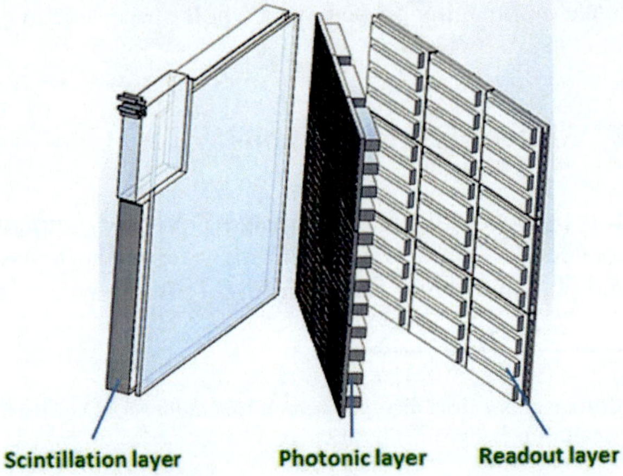

Fig.3 Layout scheme of the planar PET camera

layer composing of 31 × 31, 1.51 × 1.51 × 10 mm^3-pixel LYSO crystal array, a 144-pixels photonic layer (SensL ArrayC-30035–144-PCB), and a 1:12 linear weighting readout circuit. In this study, upper-left 4 detectors of the imaging probe were served for solution study. Although in such layout, detectors seem to be independent with each other, the transparent window does spread scintillation photons to neighbor detectors, and thus forming crossover regions when detectors leaning together. Due to lack of inter-detectors information consideration, conventional Anger logic is not capable of separating crystals in the crossover regions. Instead of the Anger logic, the ratios of the 4 acquired weighted signals, i.e. x+/x− for calculating x-positions and y+/y− for calculating y-positions, from each detector are applied to decode the event positions. Due to the pre-designed signal weights for the readout, a position mapping curve can be fitted for the weighted signal ratios (i.e. x+/x− and y+/y−),as shown in Fig. 4. Such curve is applied for position decoding of event signals to event positions (crystal 2D maps). In the proposed method, weighted signal ratios are also served as a detection of crossover region event occurrence. As shown in Fig. 5, a statistical histogram of the weighted signal ratios (x+/x−) of detector 1 (upper-left corner one).

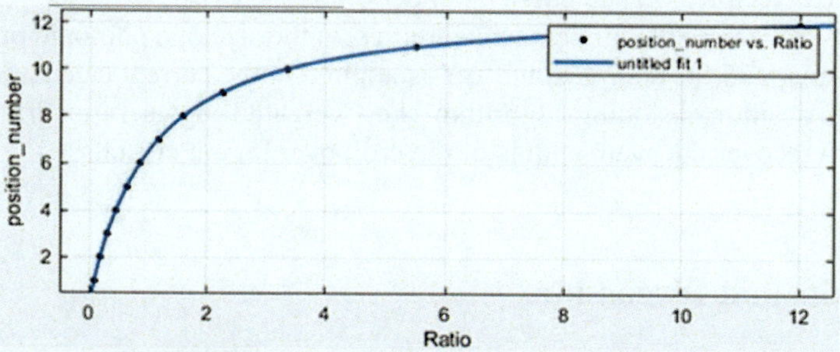

Fig. 4 Fitted curve for detector signals to find event positions

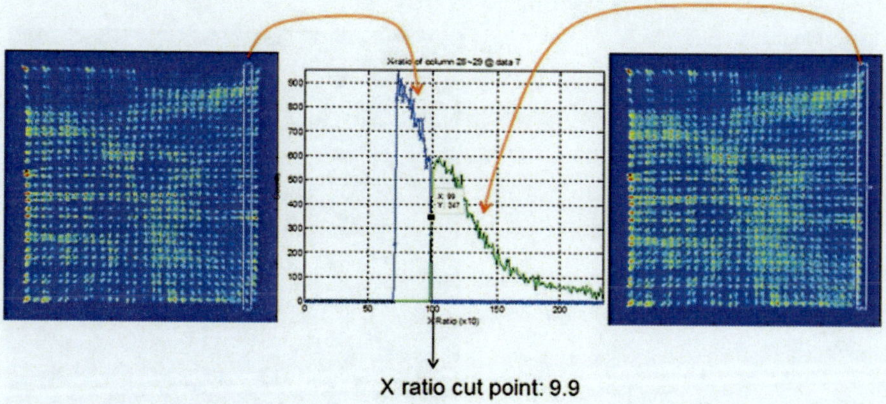

Fig. 5 A statistical histogram of the weighted signal ratios (x + /x-) of a detector

It is clearly seen that the ratio value of 9.9 can be served as a distinguishing boundary. When the weighted signal ratios approaching the ends of the mapping curve, in our case they are (larger than) 9.9 and (smaller than) 0.095, it indicates that the incident events occurred in crossover region crystals, and neighbor detector signals needed to be taken into account. For such events, position correction calculations that take neighbor detector information into account are applied, as follows:

$$X_c = X + (E_{xn}/(E_x + E_{xn}))\Delta X$$

$$Y_c = Y + \left(E_{yn}/\left(E_y + E_{yn}\right)\right)\Delta Y$$

where (Xc, Yc) are corrected event position coordinates,

(X, Y) are the uncorrected position coordinates obtained from the position mapping curve,

E_{xn} is the neighbor detector x-branch energy, i.e. $E_{xn} = (x+)_n + (x-)_n$,

E_{yn} is the neighbor detector y-branch energy, i.e. $E_{yn} = (y+)_n + (y-)_n$,

E_x is the detector x-branch energy, i.e. $E_x = (x+) + (x-)$,

E_y is the detector y-branch energy, i.e. $E_y = (y+) + (y-)$,

ΔX and ΔY are pixel numbers which represent the size of a photonic pixel in the crystal 2D map, 25 for both ΔX and ΔY is applied in the current calculations.

In such position decoding algorithm (i.e. weighted signal ratio driven), it is expected to recover the event counts of the crossover region crystals.

3 Results and Discussions

From a crystal 2D map of the proposed imaging probe exposed to a flood Cs-137 source, shown in Fig. 6, it is observed that 2 rows/columns of crystals are missing in

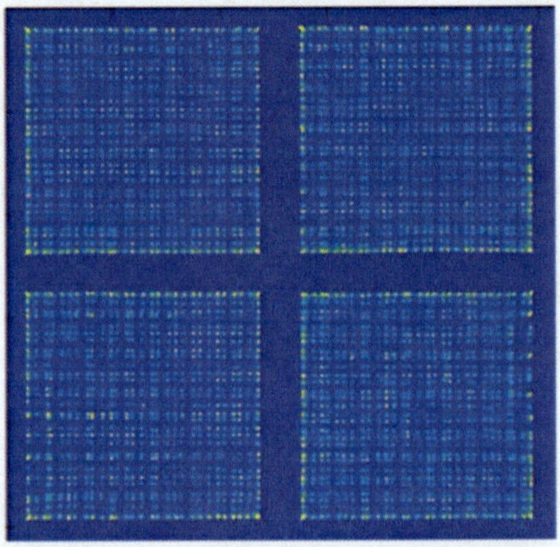

Fig. 6 Flood exposure resultant 2D map obtained with using Anger logic

Fig. 7 Crystal 2D map with applying the proposed ratio-driven position decoding. Recovering the crossover region counts is observed

the cross-over regions, and thus makes a countable crystal array of 29 × 29 for each detector, and for the upper-left 4 detectors, it is 58 × 58. Compare to the designed array size of 62 × 62, besides 2-rows/columns loss at the peripheral region, which does not affect continuity of the imaging area, the imaging probe did exhibit 2-rows/ columns loss at its center region. With applying the proposed position decoding calculation to the same imaging probe for the same experimental data, the resultant 2D map is shown in Fig. 7. It is observed that the lost 2 rows/columns of crystals in the cross-over regions are recovered, and well separations among these crystals in the 2D map are shown. It thus makes the crystal array countable element to be 30 × 30 ~ 32 × 32. And for the whole probe, it is 61 × 61. Although the loss at the peripheral region still there, the loss crystals at the probe center region are indeed recovered. It implies that the proposed new position decoding calculations work for making the imaging area our probe successive.

4 Conclusions

In this study, a set of position decoding calculations is built, and its effectiveness of making the imaging area successive is proved. The crystal array designed for our probe is 93 × 93, with the proposed calculations applying, a countable array of 91 × 91 is expected, which means the whole imaging-area of our probe is successive with the same size pixels. However, the crystals in the peripheral edges are still missing. Therefore, the following efforts will be put on searching a similar way of position decoding calculations to recover the edge rows/columns.

Acknowledgements The project budget is from AEC (Atomic Energy Commission) of the Executive Yuan. The authors would like to thank Dr. Yuan in health physics division, INER for offering and preparing flood Cs-137 radioactive sources, which are the experimental data sources of the proposed research.

Conflict of Interest The authors declare that they have no conflict of interest.

References

1. Liang HC, Jan ML, Su JL et al (2009) Development of an LYSO based gamma camera for positron and scinti-mammography. J Instrum JINST 4:P08009
2. Liang H-C, Jan M-L, Su J-L (2012) Development of a pixelated detector for clinical positron and single-photon molecular imaging. J Med Biol Eng. https://doi.org/10.5405/jmbe972

The Evaluation of Novel PET/CT Motion Correction Technique Applied in Lung Cancer Patients

Yi-Lun Chen, Bang-Hung Yang, I-Ling Shih, and Wen-Sheng Huang

1 Introduction

Recently, 18F-fluoro-2-deoxyglucose (18F-FDG) Positron Emission Tomography/Computed Tomography (PET/CT) not only offering tumor metabolism image but also morphological image simultaneously is considered a precise modality for the evaluation of lung cancer. One of the important indications is that PET/CT can effectively evaluate the malignancy of solitary pulmonary nodules and metastatic lesions. The excellent detection abilities make PET/CT scan become a clinical examination for lung cancer patients [1, 2].

However, Respiratory motion is one of the main factors significantly impacting PET/CT image quality. Motion artifacts degrade PET image quality and lead to reduction in lesion detectability, especially in lung lesions. Quantification accuracy is also impacted by respiratory motion artifacts, resulting in reduction in standard uptake value (SUV) and overestimation in lesion volume [3, 4]. These artifacts can be eliminated by 4D gated PET scan. In the past, 4D gated PET acquired data by utilizing an external gating device, e.g., elastic chest belt, Real-time Position Management (RPM) system, abdominal compression device, respiratory temperature sensor, and

Y.-L. Chen (✉) · B.-H. Yang (✉) · W.-S. Huang
Taipei Veterans General Hospital, National PET/Cyclotron Center and Department of Nuclear Medicine, Taipei, Taiwan
e-mail: ylchen47@vghtpe.gov.tw

B.-H. Yang
e-mail: bhyang@vghtpe.gov.tw

Y.-L. Chen · B.-H. Yang
Taipei Association of Radiological Technologists, Taipei, Taiwan

I.-L. Shih
GE Healthcare, Taipei, Taiwan

K.-P. Lin et al. (eds.), *Future Trends and Challenges of Molecular Imaging and AI Innovation*, Springer Proceedings in Physics 272,
https://doi.org/10.1007/978-3-030-92786-8_10

spirometer [5, 6]. Setting those external gating devices are not only time-consuming and inefficient but also increasing technologist radiation dose.

In current, a device-less data driven gated (DDG) technique (Motion free®, GE Healthcare, Waukesha, WI) was launched. It is used for deriving respiratory information from the coincidence events in each FOV. Motion is automatically detected and corrected using quiescent phase gating as the Q.Static algorithm. Then, the motion corrected images are generated. Motionfree® is an algorithm based upon principal component analysis (PCA) that determines the magnitude of respiratory impact to the data. The method analyzes the coincidence data without the need for any external device [7–10]. A respiratory waveform is derived from the PET list-mode data, and the waveform is used to derive per-cycle triggers. These triggers enable subsequent gated data processing using the existing Q.Static acquisition mode.

The Q.Static is designed to be a phase-based gating method that extracts a fraction of PET data from the end-expiration quiescent portions of patient breathing cycles. The relatively stable breathing period is acquired, in order to reduce respiratory motion. The suspicious motion area is set as the Q.Static scan mode. The respiratory motion R value was calculated in real-time. When the calculated R value was over a configured threshold, acquired time of the bed was prolonged, enabling comparable statistics as in static acquisition.

The purpose of this study was to assess that whether the novel data driven gated (DDG) technique Q.Static could improve PET respiratory motion blur in lung cancer cases.

2 Materials and Methods

A. Patient population

The current study was a retrospective study from October 2019 to February 2020. 204 patients with lung cancer were recruited. This group consisted of 105 men and 99 women with a mean age of 65.2 ± 12.6 years (range: 32–92 years) and a mean BMI of 23.7 ± 3.6 kg/m^2(range: 13.4–38.1). All patients were injected 3.7 MBq/kg $\pm$ 10% of ^{18}F-FDG after fasting a least 4 h and serum glucose concentration was less than 200 mg/dl. The PET/CT scan was obtained 60 min after injection.

B. Imaging protocol

All PET/CT scans were performed with Discovery MI DR (GE Healthcare, Waukesha, WI, USA). Acquisition protocols used clinical routine oncology whole body PET/CT protocol. Patients were requested to breathe steadily and regularly in whole body scan time. Free-breathing CT acquisition parameter was: 120 kVp, smart mA(range 30–130 mA), pitch of 0.984:1, rotation time 0.5, and 3.75 mm slice thickness (with slice increment 3.27 mm). The PET were acquired 2 min/bed from head to thigh with list-mode and time of flight (ToF) technique.

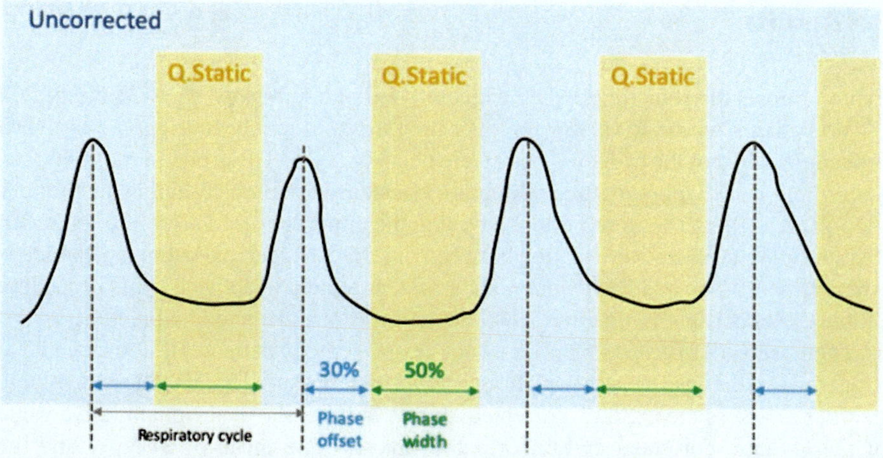

Fig. 1 Typical patient respiratory curve with a longer quiescent phase with little respiratory movement. The default Phase width is 50% and default phase offset is 30%

Data Driven Gated (DDG) Q.Static arranged to detect at Bed 3–5 which was covered whole thorax and upper abdomen. The respiratory motion R value was set 15.0 as threshold. The Q.Static parameter was modified that the phase width set 50% and phase offset set 30% which provided by the manufacturer (Fig. 1). When patients' respiratory motion R value were over than threshold, Q.Static triggered and took double acquired time to match the same image quality. At the same time, we recorded the Q.Static triggered bed for statistics.

Then we reconstructed two types of images: one was the Q.Static correction and the other was the non-Q.Static correction. Both of images used the same CT attenuation correction and the same PET reconstruction algorithm: 3-D VUE Point FX (2 iterations, 24 subsets and filter cutoff 5 mm), Sharp IR(PSF).

III. Image analysis

Image analysis was performed by AW server 3.2(GE healthcare, Waukesha, WI). Both images were drawn the 3D volume of interest(VOI) in the same size and measured the SUV_{max}、 SUV_{mean} and lesion volume. The VOI was semi-automatically drawn with a threshold 50% of SUV_{max} isocountour for each lung lesion so that reduces artificial mistakes. If the patients had more than one lesion, only the highest SUV_{max} one was included in this study.

IV. Statistical analysis

The VOI measured data were presented with mean ± standard deviation. Patients' characteristics and DDG Q.Static activation were assessed by Chi-square test. The comparison of measured data between the Q.Static corrected images and the uncorrected images was performed by Paired sample t-test. And the variation of SUV value which was corrected by the Q.Static and lesion volume were used simple linear regression analysis. P value < 0.05 was considered statistically significant.

3 Results

Table 1 shows the statistic of DDG triggered bed of all patients. The 131 out of 204 (74%) patients whose R value were over the threshold and been triggered by DDG procedure. Most of the triggered beds were bed 4 or bed 5. These beds almost covered lower lung and diaphragm area which are considered the most obvious respiratory motion area. On the contrary, Bed 3 covering the upper lung was seldom triggered in this study. We collected the relation between the DDG trigger and patients' characteristics. Patients' sex and BMI were not significant related to activate DDG procedure (Table 2). According to that, the DDG detection is only affected by breathing.

There were 131 patients mentioned above and only 60 patients' lesions located at triggered beds. These lesions were drawn VOIs and measured SUV value and volume in order to assess the Q.Static corrected ability. Table 3 shows the quantitative value of the Q.Static corrected and uncorrected images. The mean of SUV_{max} and the mean of SUV_{mean} in the Q.Static data were significantly higher than the Q.Static uncorrected data (both $P < 0.05$). Furthermore, the mean of volume in Q Static data were significantly less than uncorrected data ($P < 0.05$). The Q.Static corrected data were increased up 8.91% in SUV_{max} and 8.98% in SUV_{mean}, but decreased

Table 1 The statistic of the Q.Static triggered bed in all patients

Triggered bed	Case	Percentage (%)
Bed 3 only	0	0
Bed 4 only	37	18
Bed 5 only	45	22
Bed 3,4	0	0
Bed 3,5	0	0
Bed 4,5	46	23
Bed 3,4,5	3	1
NA	73	36
Total	204	100

Table 2 Patients' characteristic and DDG trigger

		N(%)	DDG Trigger		Significance
			yes	no	
Sex	Male	105(51.5)	67(51.1)	38(52.1)	P = 0.9
	Female	99(48.5)	64(48.9)	35(47.9)	
BMI	<18	11(5.4)	8(6.1)	3(4.1)	P = 0.196
	18–24	103(50.5)	60(45.8)	43(58.9)	
	>24	90(44.1)	63(48.1)	27(37)	

Table 3 The quantitative values and statistical significance

	SUV_{max}	SUV_{mean}	Lesion volume
Corrected images	8.77 ± 5.09	5.85 ± 3.52	4.83 ± 6.45
Uncorrected images	8.22 ± 5.04	5.47 ± 3.46	5.84 ± 7.67
Mean difference	0.54 ± 0.7	0.37 ± 0.49	−1 ± 2.11
Mean percentage difference (%)	8.91 ± 11.03	8.98 ± 10.96	−15.73 ± 20.02
Significance	P = 0.0000	P = 0.0000	P = 0.0002

down 15.73% in lesion volume. The variation of SUV_{max} and SUV_{mean} were not significantly related to lesion size ($R^2 = 0.0034$; $R^2 = 0.007$, respectively).

Figure 2 is the one interest case. According to the figure, there was one lesion in the left lower lobe. And we found that the lesion was presented blur and not concentrated in the uncorrected image. A part of lesion covered to spleen might cause a mistake. In contrast, the Q.Static corrected image showed a clear and sharp lesion. Furthermore, the quantitative data in this case was SUV_{max} increased from 8.44 to 9.82(16.35%); SUV_{mean} increased from 5.61 to 6.8(21.21%); volume reduced from 1.03 cm^3 to 0.66 cm^3(-35.92%). In other words, the Q.Static corrected image can provide better image quality and accurate quantitative data.

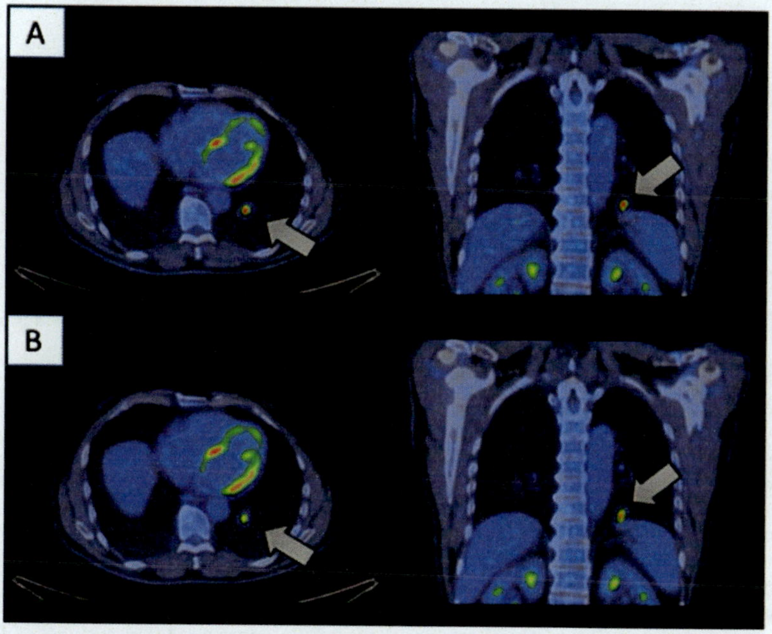

Fig. 2 A 53-year-old Male with a LLL Lung lesion. The upper row (A) was the Q.Static corrected image and the lower row (B) was uncorrected image. The arrow was shown the lesion. There were obvious blur and lengthen lesion in uncorrected image

But not all cases presented the same results. Some cases presented SUV decreasing or lesion volume increasing. We think there are two possible reasons that affect the Q.Static method. First, the lesion is located across two acquisition beds or at the overlapped area. When only one acquisition bed activates DDG, only the part of lesion is corrected and the other part is not. Thus, the quantitative values are affected. Secondly, the patient breathes irregularly and unstably. The poor breathing method has an effect on Q.Static calculation, thus causing wrong correction.

4 Conclusion

This study demonstrated that the DDG PET acquisition provides an easier way to correct motion artifacts in PET/CT scan. The quantitative SUV values increased, and lesion volumes decreased.

Acknowledgements Authors would like to express their gratitude to all patients who kindly participated in this study. And they also thank their colleague for their assistance during the study process and the imaging recruitments.

Conflict of Interest There are no conflicts of interest.

References

1. Hochhegger B, Alves GR, Irion KL et al (2015) PET/CT imaging in lung cancer: indications and findings. J Bras Pneumol 41:264–274
2. Greenspan BS (2017) Role of PET/CT for precision medicine in lung cancer: perspective of the society of nuclear medicine and molecular imaging. Transl Lung Cancer Res 6:617
3. Nehmeh SA, Erdi YE, Ling CC et al (2002) Effect of respiratory gating on quantifying PET images of lung cancer. J Nucl Med 43(7):876–881
4. Erdi YE, Nehmeh SA, Pan T et al (2004) The CT motion quantitation of lung lesions and its impact on PET-measured SUVs. J Nucl Med 45(8):1287–1292
5. Pepin A, Daouk J, Bailly P et al (2014) Management of respiratory motion in PET/computed tomography: the state of the art. Nucl Med Commun 35:113–122
6. Huang TC, Wang YC, Chou YR et al (2014) Respiratory motion reduction in PET/CT using abdominal compression for lung cancer patients. PLoS One 9(5):e98033
7. Walker MD (2019) Evaluation of data-driven respiratory gating waveforms for clinical PET imaging. EJNMMI Res 9(1):1
8. Thielemans K (2011) Device-less gating for PET/CT using PCA
9. Nehmeh SA(2004) Quantitation of respiratory motion during 4D-PET/CT acquisition. Med Phys 31(6):1333–1338
10. Walker MD, Bredley KM, McGowan DR (2018) Evaluation of principal component analysis-based data-driven respiratory gating for positron emission tomography. Br J Radiol 91(1058):20170793

Evaluation of Radiolabeling PSMA-Targeted Long Circulating Peptide as a Theranostic Agent in Human Prostate Tumor-Bearing Mice

Ming-Hsin Li, Ming-Wei Chen, Wei-Lin Lo, Yuan-Ruei Huang, Sheng-Nan Lo, Shih-Ying Lee, Su-Jung Chen, Shih-Wei Lo, Shih-Min Wang, and Chih-Hsien Chang

1 Introduction

Prostate-specific membrane antigen (PSMA) is expressed higher in prostate tumors and metastases than most normal tissue. This pathological expression pattern leads it to be a target for endoradiotherapy of prostate cancer [1, 2]. The albumin binding structure has modified to the target molecule lets the kinetic of the molecule in the blood to change. In addition, extending the circulating-time of drugs in the body is a method to enhance the efficacy of drugs [3].

In this study, we have developed a novel long-circulation PSMA-targeted molecule, MH-PC-AB-56, which PSMA binding motif (PSMA-617 structure without DOTA) covalently linked with albumin binder (4-p-tolylbutyric acid), and conjugated DOTA chelator, and labeled with radionuclide ^{111}In or ^{177}Lu. We investigated the tumor binding affinity and biodistribution by molecular imaging of radiolabeled MH-PC-AB-56 peptide. The major objective of this study was to evaluate the potential advantage of MH-PC-AB-56 in the theranostic application by NanoSPECT image.

M.-H. Li (✉) · M.-W. Chen · W.-L. Lo · Y.-R. Huang · S.-N. Lo · S.-Y. Lee · S.-J. Chen · S.-W. Lo · S.-M. Wang · C.-H. Chang
Isotope Application Division, Institute of Nuclear Energy Research, Taoyuan, Taiwan
e-mail: mhli@iner.gov.tw

C.-H. Chang
e-mail: chchang@iner.gov.tw

K.-P. Lin et al. (eds.), *Future Trends and Challenges of Molecular Imaging and AI Innovation*, Springer Proceedings in Physics 272,
https://doi.org/10.1007/978-3-030-92786-8_11

2 Materials and Methods

A. Cell culture and tumor-bearing mice

The PSMA positive prostate cancer cell line, LNCaP, was obtained from Bioresource Collection and Research Center. Cells were grown in RPMI 1640 supplemented with 10% (v/v) fetal bovine serum, 100 units/mL penicillin and 100 μg/mL streptomycin in a humidified incubator at 37 °C under 5% CO_2. Six-week-old male BALB/c nude mice were obtained from BioLASCO Taiwan Co., Ltd. Mice were housed in a 12 h light cycle at 22 °C, with food and water provided ad libitum. Cells (1×10^7) was implanted right front leg of BALB/c nude mice for LNCaP tumor xenografts. The mice underwent NanoSPECT/CT study when the tumor volume reached approximately 200–600 mm^3. The animal experimental protocols were approved by INER Institutional Animal Care and Use Committee.

B. Radiolabeling with indium-111 and lutetium-177

Indium (^{111}In) chloride in 0.01 N HCl was generated from INER (Taoyuan, Taiwan). Lutetium (^{177}Lu) chloride in 0.04 N HCl purchased from ITM Medical Isotopes GmbH (Garching/Munich, Germany). The peptide (MH-PC-AB-56 and PSMA-617) purchased from Ontores Biotechnologies Co., Ltd. (Zhejiang China).

For ^{111}In-MH-PC-AB-56 labeling, the 13.9 nmole peptide (in DMSO) was incubation with 0.23–0.26 GBq $^{111}InCl_3$ in 1 M sodium acetate solution (pH 6.0) at 95°C for 15 min. The radiochemical efficiency analyzed by thin-layer chromatography (TLC) with 10% methanol and high-performance liquid chromatography (HPLC). For ^{177}Lu-MH-PC-AB-56 labeling, the 13.9 nmole peptide (in DMSO) was incubation with 0.47–0.51 GBq $^{177}LuCl_3$ in 0.4 M sodium acetate solution (pH 5.0) at 95°C for 30 min. The radiochemical efficiency analyzed by TLC with 0.1 M citric acid and HPLC. For ^{177}Lu-PSMA-617 labeling, the 19.2 nmole peptide (in water) was incubation with 0.47–0.51 GBq $^{177}LuCl_3$ in 0.4 M sodium acetate solution (pH 5.0) at 95°C for 30 min. The radiochemical efficiency analyzed by TLC with 0.1 M citric acid and HPLC. The specific activity for ^{111}In-MH-PC-AB-56, ^{177}Lu-MH-PC-AB-56 and ^{177}Lu-PSMA-617 in the range of 16.5–18.7 GBq/μmole, 33.8–36.7 GBq/μmole and 24.5–26.6 GBq/μmole, respectively.

The radiolabeled peptides were measured using a radioactive scanner (AR-2000 radio-TLC Imaging Scanner, Bioscan, France) and radio-HPLC with UV detector (280 nm) and radio detector. The column was waters T3 C18 column (3.5 μm, 80 Å, 4.6 × 250 mm). The flow rate was 0.8 mL/min with the gradient mobile phase going from 80% A buffer (0.1% TFA in water) and 20% B buffer (0.1% TFA in acetonitrile) to 90% B buffer within 10 min for ^{111}In-MH-PC-AB-56. The flow rate was 1.0 mL/min with the gradient mobile phase going from 95% A buffer (0.1% TFA in water) and 5% B buffer (0.1% TFA in acetonitrile) to 95% B buffer within 10 min for ^{177}Lu-MH-PC-AB-56. The flow rate was 1.0 mL/min with the gradient mobile phase going from 80% A buffer (0.1% TFA in water) and 20% B buffer (0.1% TFA in acetonitrile) to 60% B buffer within 10 min for ^{177}Lu-PSMA-617.

C. In vitro stability study

The stability of ^{111}In-MH-PC-AB-56 evaluated by incubation with normal saline (volume ratio 1:1) at room temperature or human serum (volume ratio 1:19) at 37°C [4]. The radiochemical purity determined by TLC and HPLC analysis at desired times (0, 1, 4, 24, 48, 72 and 96 h). The serum sample was prepared with acetonitrile/water solution and centrifuged at 13,000 rpm 2 min, then analyzed the supernatant by HPLC.

The stability of ^{177}Lu-MH-PC-AB-56 evaluated by incubation with normal saline (volume ratio 1:1) at room temperature. The radiochemical purity determined by HPLC analysis at desired times (0, 1, 2, 3, 4, 5, 6, 7, 8 and 24 h).

D. NanoSPECT/CT imaging

The procedure for NanoSPECT/CT imaging has been described, previously [5]. NanoSPECT and X-ray images were acquired using a NanoSPECT/CT plus scanner system (Mediso Medical Imaging Systems; Budapest, Hungary). The mice was anesthetized with 1–2% isoflurane during the imaging acquisition. NanoSPECT imaging was acquired using nine multipinholes gamma-detectors and high-resolution collimators. The energy window was set at 171 and 245 keV $\pm$ 10%, the image size was set at 256 $\pm$ 256, and the field of view of 60 mm $\times$ 100 mm. Each mouse was tail-vein injected with 19.4–20.7 MBq of radiolabeled peptide and scanned for 1 h, 4 h, 24 h, 48 h, 72 h and 96 h.

For calculating uptake value at an interesting organ of ^{111}In-MH-PC-AB-56 and ^{177}Lu-MH-PC-AB-56, the images of SPECT reconstructed by HiSPECT NG software (Scivis GmbH, Germany) and fused with CT datasets using InVivoScope software (Bioscan Inc.). Then, all data processed by PMOD Version 3.3 (PMOD Technologies Ltd., Zurich, Switzerland). Volumes of Interest (VOIs) drawn encompassing the tumor and reference source on the corresponding CT images. The VOIs transferred to SPECT images, and the count values of the tumor and reference source derived. The radioactivity of reference sources were 0.8 MBq ^{111}In and 1.3 MBq ^{177}Lu.

E. Statistics

Results presented as mean and standard deviation (Mean $\pm$ SD). The distribution correlation between ^{111}In-MH-PC-AB-56 and ^{177}Lu-MH-PC-AB-56 was analyzed by Pearson correlation coefficient (r) using IBM® SPSS® Statistics software version 19 [5].

3 Results

A. Radiolabeling efficiency of MH-PC-AB-56

The labeling efficiency of ^{111}In-MH-PC-AB-56 was 98.27 $\pm$ 1.29% and 99.09 $\pm$ 0.38% by TLC and HPLC analysis, respectively. In TLC, free ^{111}In was at origin front (Rf. approximately 0.000) and ^{111}In-MH-PC-AB-56 was at solvent front (Rf.

approximately 0.917) with 10% methanol as running buffer. The retention time of ^{111}In-MH-PC-AB-56 was 8.569 ± 0.010 min in radio detector by HPLC.

The labeling efficiency of ^{177}Lu-MH-PC-AB-56 was over 99.99% and 98.87 ± 0.73% by TLC and HPLC analysis, respectively. In TLC, free ^{177}Lu was at origin front (Rf. approximately 1.023) and ^{177}Lu-MH-PC-AB-56 was at solvent front (Rf. approximately 0.015) with 0.1 M citric acid as running buffer. The retention time of ^{177}Lu-MH-PC-AB-56 was 8.587 ± 0.262 min in radio detector by HPLC.

The labeling efficiency of ^{177}Lu-PSMA-617 was over 99.99% and 98.68% by TLC and HPLC analysis, respectively. In TLC, free ^{177}Lu was at solvent front (Rf. approximately 1.045) and ^{177}Lu-PSMA-617 had Rf. 0.435 with 0.1 M citric acid as eluted buffer. The retention time of ^{177}Lu-PSMA-617 was 8.247 min in radio detector HPLC.

B. In vitro stability study of ^{111}In-MH-PC-AB-56

The radio-purity of ^{111}In-MH-PC-AB-56 was greater than 95% within 96 h incubation in normal saline or human serum (Table 1).

C. In vitro stability of ^{177}Lu-MH-PC-AB-56 in normal saline

The stability of ^{177}Lu-MH-PC-AB-56 in normal saline was greater than 95% within 8 h incubation by HPLC (Table 2).

D. NanoSPECT/CT imaging

The images showed that both radiolabeled ^{111}In/^{177}Lu-MH-PC-AB-56 and ^{177}Lu-PSMA-617 were excreted mainly via the renal pathway with higher renal retention especially at early time points 1 h and 4 h injection. Higher and sustained tumor uptake was observed for radiolabeled MH-PC-AB-56 up to 96 h by imaging (Fig. 1).

The accumulated activity of radiolabeled ^{111}In/^{177}Lu-MH-PC-AB-56 in the tumor at time points was calculated from the images created by drawing the Volume of

Table 1 In vitro stability of ^{111}In-MH-PC-AB-56 at different times after incubation in normal saline at room temperature or human serum at 37°C

Time (h)	Normal Saline		Human Serum	
	TLC (%)	HPLC (%)	TLC (%)	HPLC (%)
0	99.5 ± 0.9	98.8 ± 1.1	98.3 ± 0.8	98.8 ± 0.3
1	98.6 ± 1.3	97.9 ± 1.2	98.2 ± 0.9	98.9 ± 0.6
4	98.7 ± 1.2	97.0 ± 0.8	98.8 ± 1.1	99.2 ± 0.1
24	99.2 ± 0.7	97.8 ± 0.8	98.0 ± 1.0	99.0 ± 0.4
48	99.6 ± 0.8	97.4 ± 1.1	98.3 ± 0.6	99.0 ± 0.3
72	99.0 ± 0.8	96.6 ± 0.7	98.1 ± 0.5	98.7 ± 0.6
96	96.9 ± 2.8	98.5 ± 0.9	95.5 ± 0.8	98.8 ± 0.3

(mean ± SD, n = 3)

Table 2 In vitro stability of ^{177}Lu-MH-PC-AB-56 at different times after incubation in normal saline at room temperature (mean ± SD, n = 3)

Time (h)	Normal Saline HPLC (%)
0	99.5 ± 0.6
1	98.0 ± 0.9
2	96.6 ± 2.7
3	96.9 ± 1.0
4	94.5 ± 2.5
5	96.2 ± 1.4
6	94.9 ± 1.8
7	94.3 ± 1.1
8	94.4 ± 1.2

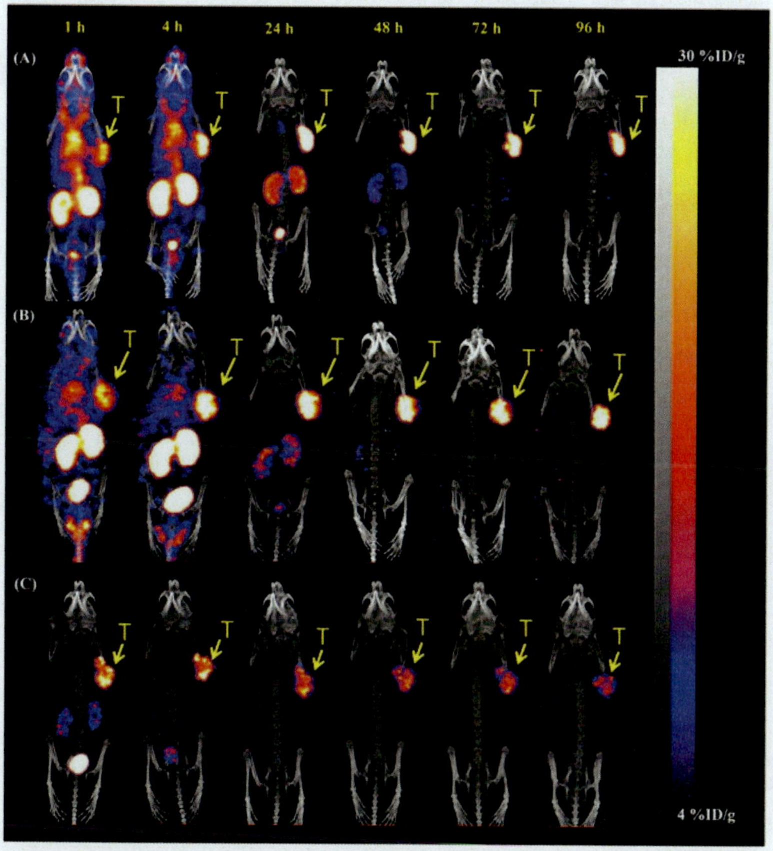

Fig.1 Comparison of NanoSPECT/CT images of ^{111}In-MH-PC-AB-56 (**a**), ^{177}Lu-MH-PC-AB-56 (**b**) or ^{177}Lu-PSMA-617 (**c**) targeting LNCaP tumor-bearing mice. The images were acquired at 1, 4, 24, 48, 72, and 96 h after injection

Table 3 Comparison of tumor uptake of ^{111}In-MH-PC-AB-56 and ^{177}Lu-MH-PC-AB-56 analyzed by NanoSPECT image

Time (h)	^{111}In-MH-PC-AB-56 (%ID/g)	^{177}Lu-MH-PC-AB-56 (%ID/g)
1	19.79 ± 2.27	12.91 ± 2.69
4	27.32 ± 2.85	17.98 ± 3.20
24	43.79 ± 6.55	22.08 ± 6.63
48	35.48 ± 4.53	21.56 ± 4.90
72	29.77 ± 4.30	22.80 ± 5.29
96	25.06 ± 3.77	23.41 ± 5.18

Interest (VOI) using the standard source as a point of reference. The image semi-quantitative analysis of ^{111}In-MH-PC-AB-56 in the tumor at 1 h, 4 h, 24 h, 48 h, 72 h and 96 h after injection was 19.79 ± 2.27, 27.32 ± 2.85, 43.79 ± 6.55, 35.48 ± 4.53, 27.77 ± 4.30, and 25.06 ± 3.77%ID/g, respectively (Table 3). The image semi-quantitative analysis of ^{177}Lu-MH-PC-AB-56 in the tumor at 1 h, 4 h, 24 h, 48 h, 72 h and 96 h after injection was 12.91 ± 2.69, 17.98 ± 3.20, 22.08 ± 6.63, 21.56 ± 4.90, 22.80 ± 5.29, and 23.41 ± 5.18%ID/g, respectively (Table 3). The trend of tumor uptake of ^{111}In-MH-PC-AB-56 is similar to ^{177}Lu-MH-PC-AB-56 before 24 h injection. After 24 h injection, the tumor uptake of ^{111}In-MH-PC-AB-56 was slightly declining, but the tumor uptake of ^{177}Lu-MH-PC-AB-56 sustained last for at least 96 h.

Pearson correlation analysis of tumor uptake of ^{111}In-MH-PC-AB-56 and ^{177}Lu-MH-PC-AB-56 by semi-quantification image analysis was moderated degree ($r = 0.576$). There was a high correlation of liver, kidney, and muscle of ^{111}In-MH-PC-AB-56 and ^{177}Lu-MH-PC-AB-56 with value 0.960 (Table 4), 0.960 (Table 5), and 0.940 (Table 6), respectively.

Table 4 Comparison of liver accumulation of ^{111}In-MH-PC-AB-56 and ^{177}Lu-MH-PC-AB-56 analyzed by NanoSPECT image (mean ± SD, n = 3)

Time (h)	^{111}In-MH-PC-AB-5 (%ID/g)	^{177}Lu-MH-PC-AB-56 (%ID/g)
1	9.89 ± 2.25	4.38 ± 1.22
4	8.31 ± 2.10	2.33 ± 0.51
24	1.84 ± 0.47	0.36 ± 0.11
48	0.53 ± 0.20	0.25 ± 0.09
72	0.32 ± 0.16	0.17 ± 0.05
96	0.23 ± 0.10	0.19 ± 0.02

Table 5 Comparison of kidney accumulation of ^{111}In-MH-PC-AB-56 and ^{177}Lu-MH-PC-AB-56 analyzed by NanoSPECT image (mean ± SD, n = 3)

Time (h)	^{111}In-MH-PC-AB-56 (%ID/g)	^{177}Lu-MH-PC-AB-56 (%ID/g)
1	37.62 ± 3.69	40.35 ± 4.89
4	39.66 ± 7.59	42.63 ± 20.14
24	16.40 ± 3.82	6.33 ± 3.51
48	6.15 ± 1.39	2.46 ± 1.39
72	3.19 ± 0.75	1.57 ± 0.95
96	2.02 ± 0.74	0.92 ± 0.68

Table 6 Comparison of muscle accumulation of ^{111}In-MH-PC-AB-56 and ^{177}Lu-MH-PC-AB-56 analyzed by NanoSPECT image (mean ± SD, n = 3)

Time (h)	^{111}In-MH-PC-AB-56 (%ID/g)	^{177}Lu-MH-PC-AB-56 (%ID/g)
1	6.58 ± 0.59	3.76 ± 1.15
4	6.70 ± 0.45	2.22 ± 0.84
24	1.29 ± 0.27	0.43 ± 0.23
48	0.41 ± 0.06	0.25 ± 0.07
72	0.21 ± 0.05	0.18 ± 0.07
96	0.17 ± 0.01	0.18 ± 0.01

4 Conclusions and Discussion

Many albumin binders (truncated Evan's blue, 4-(phenyl)butyric acid, 4-(p-bromophenyl)butyric acid, 4-(p-iodophenyl)butyric acid, 4-(p-chlorophenyl)butyric acid, 4-(p-fluorophenyl)butyric acid, 4-(p-methyl phenyl)butyric acid, and 4-(p-methoxyphenyl)butyric acid) had been used in modified PSMA-617 to enhance the circulating-time of PSMA-targeted drug to drawdown the accumulation of drug at kidney and improved tumor uptake [6]. We explored a novel linker lysine and 6-aminohexanoic acid not like PSMA-ALB-56 [7] only lysine or RPS [8] used PEG before. The structure is not only longer of length than lysine but also more lipophilic than PEG. In vitro stability study showed, the radiochemical purity of ^{177}Lu-MH-PC-AB-56 maintain to over 90% last at least 8 h without ascorbic acid addition. The SPECT image show that MH-PC-AB-56 labeled with dual-nuclide have a similar trend of metabolism or elimination organ, and demonstrated ^{177}Lu-MH-PC-AB-56 could improve tumor uptake than 177Lu-PSMA-617. These results let us know the tumor accumulated ability and elimination route of radiolabeled MH-PC-AB-56. These results indicated the ^{111}In/^{177}Lu-MH-PC-AB-56 owned the theranostic potency for diagnosis and therapy. The therapeutic efficacy, pharmacokinetics, and biodistribution would be further studied in the near future.

Acknowledgements We thanked the support from the Ministry of Economic Affairs, ROC with grant number 108-EC-17-A-22-1587.

Conflict of Interest The authors declare that they have no conflict of interest.

References

1. Rahbar et al (2018) Mol Imaging 17:1–9. https://doi.org/10.1177/1536012118776068
2. Langbein et al (2019) J Nucl Med 60:13S-19S. https://doi.org/10.2967/jnumed.118.220566
3. Kuo et al (2018) Mol Pharmaceutics 15:5183–5191. https://doi.org/10.1021/acs.molpharmaceut.8b00720
4. Lo et al (2020) Appl Radiat Isot 161:109126. https://doi.org/10.1016/j.apradiso.2020.109162
5. Chang et al (2007) Anticancer Res 27:2217–2226
6. Kuo et al (2020). J Nucl Med. https://doi.org/10.2967/jnumed.120.250738
7. Umbricht et al (2018) Mol Pharm 15:2297–2306. https://doi.org/10.1021/acs.molpharmaceut.8b00152
8. Kelly et al (2019) J Nucl Med 65:656–663. https://doi.org/10.2967/jnumed.118.221150

446

s

OK

Springer Theses

Recognizing Outstanding Ph.D. Research

Aims and Scope

The series "Springer Theses" brings together a selection of the very best Ph.D. theses from around the world and across the physical sciences. Nominated and endorsed by two recognized specialists, each published volume has been selected for its scientific excellence and the high impact of its contents for the pertinent field of research. For greater accessibility to non-specialists, the published versions include an extended introduction, as well as a foreword by the student's supervisor explaining the special relevance of the work for the field. As a whole, the series will provide a valuable resource both for newcomers to the research fields described, and for other scientists seeking detailed background information on special questions. Finally, it provides an accredited documentation of the valuable contributions made by today's younger generation of scientists.

Indexed by zbMATH.

More information about this series at http://www.springer.com/series/8790

Joseph O'Leary

General Relativistic and Post-Newtonian Dynamics for Near-Earth Objects and Solar System Bodies

Doctoral Thesis accepted by
University of South Australia, Adelaide, Australia

Springer

Author
Dr. Joseph O'Leary
Astrodynamics Group
EOS Space Systems Pty. Ltd.
Adelaide, SA, Australia

Supervisors
Prof. James M. Hill
School of Information Technology
and Mathematical Sciences
The University of South Australia
Adelaide, Australia

Associate Supervisors
Dr. James C. Bennett
EOS Space Systems Pty. Ltd.
Canberra, ACT, Australia

Prof. Sam Drake
College of Science and Engineering
Flinders University
Adelaide, Australia

ISSN 2190-5053 ISSN 2190-5061 (electronic)
Springer Theses
ISBN 978-3-030-80184-7 ISBN 978-3-030-80185-4 (eBook)
https://doi.org/10.1007/978-3-030-80185-4

This Springer imprint is published by the registered company Springer Nature Switzerland AG
The registered company address is: Gewerbestrasse 11, 6330 Cham, Switzerland

Supervisor's Foreword

The next generation of near-Earth satellite systems will require a detailed mapping of the Earth's gravitational field combined with the development of accurate predictive mathematical modelling. Accordingly, future space scientists will be required to have an understanding and knowledge of practical general relativity, as suggested by the International Astronomical Union, that all astronomical problems should be formulated within the framework of general relativity. Hence, all future mission planning tools and orbit determination software suites must address this issue. Astronomical problems posed within the framework of general relativity present a formidable task both in terms of shear complexity and depth of the gravitational field equations and do not submit to a comparable analytical treatment as their classical Newtonian counterparts. This text is designed for those future space scientists requiring a working knowledge of the practical aspects of general relativity.

Einstein's general theory of relativity was published over a century ago and provides an accurate description of gravitation giving rise to a myriad of interesting and peculiar phenomena in our universe, such as the bending of light through gravitational lensing, the slowing of clocks in gravitational fields and the recently detected ripples in spacetime due to cataclysmic astrophysical events such as the coalescence of dense stellar objects.

The practical importance of general relativity is evident for the description of exotic compact astronomical objects such as neutron stars and black holes, for which classical Newtonian gravity does not capture the qualitative behaviour of such strong-field phenomena. While fundamental in fields such as cosmology and gravitational wave astronomy; general relativity might be perceived by some as somewhat esoteric and academic when describing the interaction of bodies in the weak-field arena of the solar system. However, this could not be further from the truth which is the central theme of this book.

Given the increased accuracy requirements in fields such as astrometry and geodesy, and owing to the rapid improvements in time and frequency technology, Einstein's general theory of relativity can take such matters into account for those space missions requiring highly accurate orbit information and practically all observation techniques. The general relativistic problems in this book have both theoretical and practical applications and demonstrate the confluence of applied mathematics, general relativity and space engineering through examination of problems of space science.

The first part of the book begins with the exact vacuum solution to the Einstein field equations of gravitation by Karl Schwarzschild which was discovered shortly after the publication of Einstein's general relativity. Birkhoff's theorem states that all spherically symmetric solutions are static and asymptotically flat, which in consequence, means that any vacuum solution exhibiting spherical symmetry is equivalent to that of Schwarzschild. While this is known to practitioners of general relativity, the Schwarzschild solution contains coordinate singularities and requires additional transformations to achieve maximal analytic coverage of the underlying spacetime. Chapter 2 explores Birkhoff's theorem and derives an alternative, singularity-free solution to the Einstein field equation exhibiting an additional degree of freedom, by directly solving the resulting system of non-linear differential equations.

The second part of this book is dedicated to the gravitational interaction between objects in the solar system using the so-called post-Newtonian approximation. The post-Newtonian regime has been coined *unreasonably effective* in describing gravitational physics and is an invaluable tool in modern astrodynamics and geodesy. Following an introduction to the post-Newtonian framework in Chap. 3, the author explores three post-Newtonian dynamical problems, namely: energy integrals for spacecraft in the local geocentric frame, the post-Newtonian Kepler and two-body problems and the relativistic Euler or two-centre problem.

The derivation of the Newtonian or Jacobi energy integral is known to all students of mechanics. The post-Newtonian equivalent is presented in Chap. 4, followed by an extensive numerical investigation using sophisticated structure-preserving symplectic integration schemes. The remainder of the second part of the book explores the post-Newtonian extension of the Kepler and Euler problems, respectively. In Chap. 5, the post-Newtonian extension of the well-known Binet equation associated with the classical two-body problem is derived. Unlike its classical counterpart, the relativistic Euler problem eludes analytical integration in Chap. 6. However, the resulting system of equations of motion gives rise to a global first integral in the two-centre problem, verifying the results of Chap. 4.

The final part of the book discusses the development of a general relativistic orbit propagation tool. While the industry standard is to include general relativity as an effective force which is linearly combined with Newton's second law, a novel approach is described to tackle this problem using the full framework of general relativity in Chap. 7. The software known as General Relativistic Accelerometer-based Propagation Environment (GRAPE) presently under development shows promising results and will serve as a test-bed for current and future propagation software.

Adelaide, Australia
May 2021

Prof. James M. Hill

Preface

The first time I was introduced to Einstein's general theory of relativity was terrifying. As an undergraduate student with only a semester of classical mechanics in my problem-solving arsenal, battling with the ever-changing gymnastics of tensor indices and tackling with concepts like spacetime curvature and the existence of exotic stellar objects such as black holes was daunting to say the least. While initially apprehensive and intimidated to open a copy of Misner, Wheeler and Thorne's celebrated and dense text, the theory of relativity now provides me with endless hours of happy and confused thought.

The current body of work is the product of three very positive and productive years as a doctoral candidate with the University of South Australia under the guidance of Prof. Jim Hill, Dr. James Bennett and Professor Sam Drake. While initially interested in the theoretical aspects of general relativity, my academic journey was perturbed through my unexpected and fortuitous involvement with the Australian Space Environment Research Centre and industry partner EOS Space Systems. As a student involved in both academic and industry sectors, I was fortunate enough to see the merits in both theoretical and applied research domains which has greatly shaped my current career and research interests in the space sector.

There are many excellent texts which address popular topics in general relativity such as black holes, cosmology and gravitational waves. However, only a handful of texts spring to mind which discuss real-world applications of relativity in the weak-field domain of the solar system. Texts such as Soffel (1989); Kopeikin et al. (2011); Brumberg (2017) and most recently Soffel and Han (2019) provided much inspiration for many of the topics covered in the current book and greatly facilitated and assisted in my journey as a doctoral candidate. My goal when writing the current text was to present several problems in applied general relativity which can be understood not only by those who are well-versed in the language of general relativity, but also by applied mathematicians, space scientists and engineers.

In this book, problems from three areas of general relativity are investigated:

- Exact solutions of the Einstein field equations of gravitation.
- Dynamics of spacecraft and celestial bodies in the first post-Newtonian approximation to general relativity.
- Relativistic orbit determination of spacecraft and interplanetary probes including non-gravitational perturbations.

Part II of the book examines standard methods adopted when deriving exact, analytical and spherically symmetric solutions to the Einstein field equations. The standard approach in deriving the well-known Schwarzschild solution is to introduce convenient coordinate transformations, removing any off-diagonal metric tensor components and to adopt physical arguments which inevitably lead to the Schwarzschild solution. However, it is well known the resulting spacetime contains coordinate singularities and requires further transformations. Motivated by this, we derive an alternative spherically symmetric solution which does not contain the Schwarzschild singularity. By retaining the off-diagonal time-space components of the metric tensor, we derive the only alternative spherically symmetric solution to the field equations by directly solving the corresponding non-linear system of partial differential equations.

Part III is concerned with the well-known post-Newtonian approximation to general relativity. We consider two separate problems, namely, the relativistic two-body and two-centre Euler problems respectively. First, we derive a new energy-like integral associated with a test body. The resulting energy integral contains exponential functions, and, when linearly approximated produces results in the literature. Similarly, we derive a post-Newtonian Binet equation which characterises the shape of satellite orbits in the first post-Newtonian regime. Finally, we derive the general first integral for an arbitrary gravitational potential associated with a test particle which moves in the gravitational fields of two attracting centres.

The final part of this thesis suggests a framework for general relativistic orbit determination which includes non-gravitational perturbations. This is a problem of great interest to the community for which a suitable solution has not yet been given. Standard orbit determination software numerically integrates Newton's second law perturbed by linearly superimposed effective forces. The *effects* of general relativity are treated in the same manner using the classical Newtonian concept of absolute spatial and temporal coordinates. Here, we present a newly developed software which describes the perturbed motion of a test particle in the complete framework of general relativity.

This book is presented in six parts. Part I comprises the introduction and overview of the present body of work. Part II is concerned with the Einstein field equations of gravitation, and in particular, exact solutions for spherically symmetric, non-orthogonal line elements. In Part III, we introduce the reader to the successful Damour, Soffel and Xu post-Newtonian formalism and proceed to develop new dynamical properties of gravitating bodies in both local and global frames. Following this, Part IV suggests a new framework for general relativistic orbit determination which includes non-gravitational perturbations. We note this work is still ongoing.

Hence, this chapter is brief. Part V summarises the results of this book and Part VI comprises the relevant appendices and bibliography. Finally, we note, when appropriate, a nomenclature is provided for chapters which are considered substantial, and we relegate appropriate reviews of relevant literature to the beginning of each respective chapter.

Adelaide, Australia
May 2021

Dr. Joseph O'Leary

Acknowledgements

It is a difficult task to appropriately acknowledge all the people who made my Ph.D. so enjoyable in such a short space, and to do so adequately would require the same amount of time as writing the current body of work. First, I must thank Jim Hill for his guidance, patience and commitment to helping me over the years. The many tribulations encountered during a Ph.D. can be disconcerting and I am grateful and indebted to Jim for teaching me how to learn from each of them. Second, I must acknowledge the support from my co-supervisor and good friend Dr. James Bennett, who has always supported my left-of-centre ideas and never faltered when they failed. On those very rare occasions when they were successful; James was never too far away to celebrate in a manner that reflects the cultural appreciations of those from the Emerald Isle. Further, I must acknowledge many helpful discussions with Dr. Sam Drake. Throughout my Ph.D., Sam has always provided me with the "physicists way of thinking" which has aided many chapters of this thesis. Last but not least, I'd like to thank Dr. Barry Cox, who never refused a coffee invitation; even if I would pester him with questions.

I've enjoyed overseas conferences, seminars and technical workshops which would not have been possible without the financial support from the Space Environment Research Centre and the University of South Australia. I would like to mention Steve Gower, Michelle Fulton, Michael Lachut, Sven Flegel, Daniel Kucharski, Marek Moeckel, David Ball, Craig Smith and Ben Greene who have provided support; answers to questions I probably should know; and patience when I proceeded to say I still didn't understand. Further, I would like to acknowledge Prof. Jean-Pierre Barriot from the University of French Polynesia who provided financial support to visit Tahiti and the idea for the content of Chap. 7.

On a personal level, I owe a debt of gratitude to my sisters, Stacey and Lynsey, for their constant support and encouragement. To my mam, Jackie and to Deirdre, you have the innate ability to put a positive spin on things; regardless of how glum they first appear. Finally, I would like to give my sincerest and heartfelt thanks to my partner Enya. While moving interstate is an incredible and selfless act of support, Enya always listened to my strange ideas and offered feedback, watched all my boring presentations with excitement and read several drafts of my thesis without batting an eyelid. For this, I will always be grateful.

Contents

List of Figures

Part I
Introduction

Chapter 1
Overview

The general theory of relativity, proposed by Albert Einstein over a century ago, provides our currently accepted description of gravitation. When first proposed, the validity of the framework rested on three famous tests:

- the theoretical accounting for the relativistic contribution to the perihelion precession of Mercury.
- Observing the deflection of light around a massive body.
- The measurement of gravitational redshift.

Other early experimental tests were performed by Hafele and Keating (1972) and Allan et al. (1985), who compared theoretical predictions of general and special relativistic time dilation using airborne atomic clocks and electromagnetic signals from the global positioning system respectively. Since then, and with the rapid advancement in fields such as optical and quantum sciences, time dilation has been measured with relative motion as low as 10 m/s and observed for height differences of 1 m in the gravitational field of the Earth (Chou et al. 2010).

The initial experimental tests of general relativity appear mundane when compared with modern and sophisticated tests such as the detection of gravitational waves. First predicted by Einstein (1916a, b, 1918), gravitational waves were initially detected by the advanced laser interferometer gravitational-wave observatory collaboration (Abbott et al. 2016). Since then, several gravitational waves have been detected including the simultaneous detection of gravitational waves and electromagnetic radiation due to the merging of a binary neutron star system (Goldstein et al. 2017). More recently, further verification for the general theory of relativity was provided by the Event Horizon Telescope Collaboration (2019) who obtained the first image of the shadow of a black hole. The once invisible arena in which gravity dominates is now visible (see Barack et al. (2019) for a thorough and concise account on gravitational wave astronomy).

J. O'Leary, *General Relativistic and Post-Newtonian Dynamics for Near-Earth Objects and Solar System Bodies*, Springer Theses,
https://doi.org/10.1007/978-3-030-80185-4_1

It is often suggested that the classical theory of Newtonian gravity was supplanted by the general theory of relativity. However, this is misleading. In fact, general relativity provided the first acceptable answer as to *why* Newtonian gravity was so successful at describing gravitational phenomena within the solar system (Hawking and Israel 1989, Chap. 2). It was observed very early in the development of the general theory of relativity that one can recover Newtonian gravity under certain dynamical assumptions. Under such assumptions, the recovery of the well-known inverse square-law of Newtonian theory is unavoidable, which, until the inception of general relativity was based purely on empirical evidence.

1.1 Exact Solutions to the Field Equations of Gravitation

When the gravitational field is assumed everywhere weak and the bodies move slowly when compared with speed of light, the Einstein field equations reduce to that of Newton; this is known as the Newtonian limit. However, when these assumptions are not made, the Einstein field equations (Stephani et al. 2009) consist of a system of ten non-linear, coupled partial differential equations. Due to the non-linearity of the Einstein field equations, the process of finding exact, analytical solutions presents a formidable task to mathematicians and physicists alike. When combined with the requirement that an exact solution have realistic astrophysical applications; the problem quickly becomes intractable. Notable vacuum solutions to the Einstein field equations are given by Schwarzschild (1916) and Kerr (1963) which describe isolated, non-rotating, spherically symmetric and rotating, axially symmetric black holes respectively.

The derivations of the Schwarzschild solution are common in undergraduate literature (d'Inverno 1992; Misner et al. 1973), and due to the so-called Birkhoff theorem (Hawking and Ellis 1973), vary little from reference to reference. The approach employed in this thesis bifurcates from the standard derivation approach. From the outset, we assume a non-orthogonal spacetime line element with off-diagonal, spatially and temporally dependent metric tensor components. The resulting non-linear system of partial differential equations is systematically solved which not only confirms Birkhoff's theorem but also provides an alternative spherically symmetric solution to that of Schwarzschild.

1.2 Post-Newtonian Dynamics and Celestial Mechanics

The next level of approximation is known as the first post-Newtonian approximation which leads to observable relativistic phenomena such as the relativistic contribution to planetary perihelion precession. The post-Newtonian framework (Damour et al. 1991; Kopeikin et al. 2011; Soffel 2014; Poisson and Will 2014) is an important tool for the successful accounting of general relativity within the solar system and is relied on heavily in fields such as fields such as geodesy (Müller et al. 2008; Puetzfeld and Lämmerzahl 2019) and interplanetary navigation (Moyer 1971, 2005). Further, it is recommended by the international astronomical union that fields such as astrometry, celestial mechanics and metrology be defined in the framework of general relativity which is achieved at the first post-Newtonian level.

It has been observed by the author, that the theoretical aspect of the post-Newtonian regime has not received the same exhaustive treatment as classical Newtonian dynamics (Whittaker 1988) and many problems require generalisations. In this thesis, we investigate three areas of the Damour, Soffel, Xu post-Newtonian approximation (Damour et al. 1991, 1992, 1993, 1994). Firstly, we consider the first integral for energy associated with a test particle in the local frame of the Earth. Second, we derive the analytical post-Newtonian Binet equation in the centre of mass frame for the relativistic Kepler and one-body problems respectively. Finally, we generalise the two-centre Euler problem (Ó Mathúna 2008) for first post-Newtonian dynamics and derive the corresponding, non-separable first integral.

1.3 General Relativistic Orbit Determination

In modern planetary orbitography and spacecraft orbit determination software, the motion of objects is still described using Newton's second law of motion with gravitational and non-gravitational perturbations being linearly superimposed as required. Past a certain threshold and with the precision of measurement technology increasing, general relativity must be accounted for. The recommended equations of motion are provided through technical documentation maintained by the international earth rotation and references systems services (Petit and Luzum 2010). The corresponding framework for orbit determination in general relativity does not exist.

To accurately determine orbits using general relativity, we require that external perturbations be expressed in a Lorentz-invariant form with a corresponding component in the temporal direction. Currently perturbation models are only given in the absolute spatial and temporal coordinates of Newtonian gravity (Roh et al. 2016). In this thesis, we generalise the work of Pireaux et al. (2006) to integrate the full general relativistic, four-dimensional equations of motion in the Schwarzschild field of the Earth using the NASA Parker solar probe as a test body. The approach adopted here relies on the formalism presented in Lichnerowicz and Teichmann (1955) where the non-gravitational perturbations are encoded in a four-vector which materialises on the right-hand side of the standard geodesic equation of motion. In order to determine the spatial components of the perturbations, we employ on-board accelerometers which measure non-gravitational forces to high accuracy. The temporal component can then be determined using the constancy of the magnitude of the four-velocity in the local frame of the spacecraft. The approach suggested here is intended to serve as a highly accurate test bed for future orbitography and orbit determination software.

Part II
The Field Equations of General Relativity

Chapter 2
Spherically Symmetric General Relativity

2.1 Introduction

In this chapter, we derive a new, static, spherically symmetric solution to the empty space field equations of general relativity. The new solution arises by studying the system of coupled, non-linear partial differential equations associated with the non-orthogonal invariant line element (2.4) and contains two arbitrary constants of integration. Using the weak-field limit of general relativity, we identify one constant with the classical Newtonian gravitational potential, and the additional constant allows for a generalisation of the Eddington–Finkelstein transformation and the Kruskal–Szekeres coordinates.

The chapter is organised as follows: in Sect. 2.4 a description of the governing equations which leads to both solutions is presented. In Sect. 2.5, we present a novel derivation of the Schwarzschild solution and in Sect. 2.6 we derive a new spherically symmetric solution to the empty space field equations of general relativity. In Sect. 2.7 we obtain the corresponding generalisations of the Eddington–Finkelstein transformation and the Kruskal–Szekeres coordinates. In Sect. 2.8, we present the geodesic equations of motion associated with a test particle orbiting a central body whose spacetime geometry is described by (2.24) and derive the corresponding general relativistic Binet equation. Finally, the non-zero, independent Christoffel symbols and mixed and covariant Einstein tensor components associated with Eq. (2.4) are presented in Appendix A.

The following chapter and some of the results discussed have been published in Hill and O'Leary (2018)

- Hill, J.M. and O'Leary, J., 2018. Generalised transformations and coordinates for static spherically symmetric general relativity. *Royal Society open science*, 5(4), p.171109

J. O'Leary, *General Relativistic and Post-Newtonian Dynamics for Near-Earth Objects and Solar System Bodies*, Springer Theses,
https://doi.org/10.1007/978-3-030-80185-4_2

2.2 Nomenclature

M	is the mass of a central gravitating body
r_0	is the Schwarzschild radius
c	is the speed of light in vacuum
G	is the Newtonian gravitational constant
$G_{\alpha\beta}$	are the components of the covariant Einstein tensor
$R_{\alpha\beta}$	are the components of the covariant Ricci tensor
R	is the Ricci scalar
$g_{\alpha\beta}$	are the metric tensor components
$\mathcal{L}$	is the general relativistic Lagrangian function
$\mathcal{H}$	is the general relativistic Hamiltonian function
$\Gamma^{\alpha}_{\beta\gamma}$	are the Christoffel symbols of the second kind
$\mathcal{J}$	is the angular momentum per unit mass
$\mathcal{E}$	is the test particle energy per unit mass
$\mathcal{R}, \mathcal{T}$	are the generalised Kruskal–Szekeres coordinates
$\mathcal{R}, \mathcal{T}$	are the standard Kruskal–Szekeres coordinates.

2.3 Background and Motivation

The first exact, analytical solution to the vacuum field equations of GR is due to Karl Schwarzschild (Schwarzschild 1916). The derivation is now commonplace and can be readily found in the literature (Dalarsson and Dalarsson 2015; d'Inverno 1992; Foster and Nightingale 2010; Misner et al. 1973; Ohanian and Ruffini 2013; Schutz 2009). The Schwarzschild solution describes a spherically symmetric, static and asymptotically flat spacetime outside a body of mass M. The line element in Schwarzschild geometry in spherical polar coordinates $x^{\mu} = (ct, r, \theta, \varphi)$ is given by

$$\mathrm{d}s^2 = (1 + r_0/r)c^2\mathrm{d}t^2 - (1 + r_0/r)^{-1}\mathrm{d}r^2 - r^2\mathrm{d}\Omega^2, \tag{2.1}$$

where the line element of a two-sphere, the Schwarzschild radius, the speed of light in vacuum and Newton's gravitational constant are given by $\mathrm{d}\Omega^2 = \mathrm{d}\theta^2 + \sin^2\theta\mathrm{d}\varphi^2$, $r_0 = -2GM/c^2$, c and G respectively. In what follows and for the remainder of this chapter, we assume a time-scale for which $c = 1$.

The standard approach in obtaining the Schwarzschild solution is as follows:

1. As an ansatz, choose the most general spherically symmetric line element given by (Cheng 2005)

$$\mathrm{d}s^2 = a(r,t)\mathrm{d}t^2 + 2b(r,t)\mathrm{d}r\mathrm{d}t - b(r,t)\mathrm{d}r^2 - d(r,t)r^2\mathrm{d}\Omega^2, \tag{2.2}$$

where a, b, c and d denote unknown functions of both space and time which are to be determined.

2. Assume the spacetime is static such that the metric tensor components are time independent *and* time reversible (Hobson et al. 2006).
3. Introduce a new radial coordinate $r' = r\sqrt{d(r)}$ to remove the unknown function $d(r)$.
4. Remove the off-diagonal metric tensor component $b(r)$ associated with the cross term $\mathrm{d}r\mathrm{d}t$ which is achieved by introducing a new temporal coordinate t' through the differential relations

$$\mathrm{d}t' = \mathrm{d}t + \frac{\mathrm{d}f}{\mathrm{d}r}\mathrm{d}r, \qquad (\mathrm{d}t')^2 = \mathrm{d}t^2 + 2\frac{\mathrm{d}f}{\mathrm{d}r}\mathrm{d}r\mathrm{d}t + \left(\frac{\mathrm{d}f}{\mathrm{d}r}\right)^2 \mathrm{d}r^2.$$

Hence, the cross term $\mathrm{d}r\mathrm{d}t$ is subject to the constraint

$$\mathrm{d}r\mathrm{d}t\left(\mathrm{b}(r) - \mathrm{a}(r)\frac{\mathrm{d}f}{\mathrm{d}r}\right) = 0,$$

and is removed if the differential equation

$$\frac{\mathrm{d}f}{\mathrm{d}r} = \frac{b(r)}{a(r)},$$

is satisfied.
5. Calculate the components of the Einstein tensor $G_{\alpha\beta}$ and solve the system $G_{\alpha\beta} = 0$ for the remaining unknown functions $\mathrm{a}(r)$ and $\mathrm{c}(r)$. We note the systems $G_{\alpha\beta} = 0$ and $R_{\alpha\beta} = 0$ are equivalent in vacuum which follows from calculating the trace of the vacuum field equations

$$g^{\alpha\beta}G_{\alpha\beta} \equiv g^{\alpha\beta}R_{\alpha\beta} - \frac{1}{2}g^{\alpha\beta}g_{\alpha\beta}R = 0,$$

which implies $R = 0$ in vacuum.

The aforementioned approach then guarantees that the only possible solution to the vacuum field equations for a line element of the form

$$\mathrm{d}s^2 = a(r)\mathrm{d}t^2 - c(r)\mathrm{d}r^2 - r^2\mathrm{d}\Omega^2, \tag{2.3}$$

is given by Eq. (2.1) as inferred by the so-called Birkhoff's theorem (Abbassi 2001; Birkhoff 1923; Johansen and Ravndal 2006).

Birkhoff (1923) was the first to provide a solution to the spherically symmetric line element without assuming a static spacetime (Wald 1984), and the reader is encouraged to consult Hawking and Ellis (1973) and Weinberg (1972) for proofs.

Formally, Birkhoff's theorem states all spherically symmetric vacuum solutions to the Einstein field equations are necessarily static and any alternative spherically symmetric solutions are locally isometric to the Schwarzschild solution. A further consequence is that, although a pulsating spherically symmetric body can produce time dependent phenomena within its mass distribution, no gravitational radiation can be expelled in to empty space (Weinberg 1972) i.e. general relativity, as well as electromagnetism, does not permit monopole radiation (Wald 1984).

The present chapter does not dispute Birkhoff's theorem and is not intended as a rebuttal of its validity. In fact, the results which follow provide an independent verification of Birkhoff's theorem while also deriving an alternative spherically symmetric solution to the empty space field equations associated with the line element

$$ds^2 = a(r,t)dt^2 + 2b(r,t)drdt - c(r,t)dr^2 - r^2d\Omega^2. \tag{2.4}$$

Motivation for such a study is as follows: The Schwarzschild solution contains a coordinate singularity at $r = r_0$ and requires further transformations and extensions to exhibit non-singular structure. This suggests that the resulting line element is described by a coordinate chart that is not optimal and warrants an investigation. The key point of this chapter is that if steps (2) and (4) above are ignored, another solution opportunity presents itself which suggests an improved coordinate chart to that of Schwarzschild. Furthermore, the alternative solution allows for the generalisation of the Eddington–Finkelstein and Kruskal–Szekeres coordinate transformations. By making the usual simplifying assumptions and coordinate transformations at the outset the solution presented below is excluded.

2.4 Governing Equations

The metric tensor $g_{\alpha\beta}$ and the inverse metric tensor $g^{\alpha\beta}$ associated with (2.4) are given by

$$g_{\alpha\beta} = \begin{bmatrix} a & b & 0 & 0 \\ b & -c & 0 & 0 \\ 0 & 0 & -r^2 & 0 \\ 0 & 0 & 0 & -r^2\sin^2\theta \end{bmatrix}, \quad g^{\alpha\beta} = \begin{bmatrix} c/\omega & b/\omega & 0 & 0 \\ b/\omega & -a/\omega & 0 & 0 \\ 0 & 0 & -r^{-2} & 0 \\ 0 & 0 & 0 & -(r^2\sin^2\theta)^{-1} \end{bmatrix}, \tag{2.5}$$

where a, b and c are unknown functions of both space and time to be determined and $\omega(r,t) \equiv ac + b^2$. It is important to note that in order to obtain the inverse metric tensor $g^{\alpha\beta}$ we require $\det g_{\alpha\beta} \neq 0$ for which the condition $\omega(r,t) \neq 0$ immediately follows.

The first equation under consideration is Eq. (A.9) which is given by

$$G_{00} = \left[a\left(a^2c_r - a_rb^2 + 2abb_r\right)r + \omega^2 - a\omega\right]/(\omega r)^2 - b(a^2c_t + 2abb_t - a_tb^2)/\omega^2 r = 0. \tag{2.6}$$

Introducing subscript notation to denote a partial derivative with respect to the corresponding coordinate, Eq. (2.6) is equivalently expressed as

$$a\left[1-\frac{a}{\omega}-r\left(\frac{a}{\omega}\right)_r\right]-rb\left(\frac{a}{\omega}\right)_t=0,$$

which reduces to the simple linear partial differential equation

$$a(r,t)=au_r-bu_t, \tag{2.7}$$

where we have introduced $u(r,t)\equiv ra/\omega$. Similarly, the off diagonal component of the covariant Einstein tensor G_{01} is given by Eq. (A.10) and is expressed in simplified form according to

$$\frac{r}{\omega}(ac_t+2bb_t)-b\left(\frac{ra}{\omega}\right)_r-\frac{r\omega_t}{\omega^2}(\omega-ac)=0,$$

and produces a second linear partial differential equation given by

$$b(r,t)=bu_r+cu_t. \tag{2.8}$$

The system of partial differential equations (2.7) and (2.8) is immediately soluble, for which we may readily deduce $u_r=1$ and $u_t=0$. An expression for the unknown metric tensor component $a(r,t)$ is obtained following integration of $u_r=1$ and is given by

$$a(r,t)=\omega(r,t)\left(1+\frac{r_0}{r}\right), \tag{2.9}$$

where r_0 is an arbitrary constant of integration which we anticipate will correspond to the Schwarzschild radius. The remaining equations of interest arise from algebraic manipulation of (A.11) and (A.12) respectively.

Equation (A.11) is given by

$$\begin{aligned}G_{11}=c\left\{\left[abc_t+(aa_r+ba_t-2ab_t)c+a_rb^2\right]r-\omega^2+a\omega\right\}/(\omega r)^2\\+b\left[\omega c_t+b(ac_r+ca_r+2bb_r)\right]/\omega^2r=0,\end{aligned}$$

which, after some simplification reduces to

$$cu_r-\frac{r}{c}\left(\frac{b}{\omega}\right)_t+\frac{r}{\omega}(bc_t-cb_t)+\frac{r\omega_r}{\omega}=c.$$

Introducing expressions $v\equiv rb/\omega$, $w\equiv rc/\omega$ and dividing the above equation by c^2 yields

$$u_r-v_t=1+w\left(\frac{v}{w}\right)_t,$$

and using the identity $u_r=1$ in the above equation yields an integrable partial differential equation

$$2v_t/v=w_t/w, \tag{2.10}$$

which, upon integration, produces an expression for $b(r, t)$ given by

$$b(r, t) = \left[\frac{\omega(r, t)}{r} f(r) c(r, t) \right]^{1/2}, \tag{2.11}$$

where $f(r)$ is introduced as an arbitrary function of integration.

The final unknown metric tensor component $c(r, t)$ is obtained by substituting Eqs. (2.9) and (2.11) into $\omega(r, t) = ac + b^2$. Hence, $c(r, t)$ is given by

$$c(r, t) = \frac{r}{r + r_0 + f(r)}, \tag{2.12}$$

where it follows immediately, $c(r, t)$ is at most a function of r, so that $c(r, t) = c(r)$. Conversely, using the fact $c_t = 0$ and $b(r, t)$ is assumed to be non-zero, we deduce $\omega(r, t)$ is at most a function of time from Eq. (2.8), so that $\omega(r, t) = \omega(t)$. Further, we obtain a reduced and simplified expression for $b(r, t)$ by substituting (2.12) in (2.11) given by

$$b(r, t) = \left[\frac{\omega(t) f(r)}{r + r_0 + f(r)} \right]^{1/2}. \tag{2.13}$$

The final equation of interest is obtained by multiplying (A.12) by $-4\omega(t)^2/r$, and is given by

$$\begin{aligned}
&-4\omega^2 G_{22}/r = \{2\omega c_{tt} - a c_t^2 - (a_t c + a_r b + 2bb_t) c_t \\
&+ (a a_r + a_t b - 2ab_t) c_r + (a_r^2 - 2a a_{rr} + 4ab_{rt} - 2a_t b_r) c \\
&- 2b(a_r b_r - a_{rr} b + 2b_r b_t + 2bb_{rt})\} r - 2a(a c_r + b c_t) \\
&+ 2(2ab_t - a_t b - a a_r) c + 4b(ab_r - a_r b) = 0.
\end{aligned}$$

The simplification of the above equation is facilitated by introducing $\omega = \omega(t)$ and $c_t = 0$ and is given by

$$4\omega(rb)_{rt} - 2\omega_t (rb)_r - 2ab_t (rc)_r - 2\omega(t)(ra)_{rr} - b\omega_t - 2\omega b_t = 0.$$

Dividing throughout the above equation by $2\omega(t)^{3/2}$ we find

$$2\left(\frac{(rb)_r}{\omega(t)^{1/2}} \right)_t - \left(\frac{(ra)_r}{\omega(t)^{1/2}} \right)_r + \left(\frac{b}{\omega(t)^{1/2}} \right)_t - ab_t \left(\frac{(rc)_r}{\omega(t)^{1/2}} \right) = 0, \tag{2.14}$$

where it is clear the time dependent components in the first and third expressions in Eq. (2.14) will cancel with the denominators in each case. Further, the second term in Eq. (2.14) is the second derivative of a linear function, so the only remaining non-zero term is

$$\left(1+\frac{r_0}{r}\right)\frac{\omega_t(t)}{2}\left[\frac{f(r)}{r+r_0+f(r)}\right]^{1/2}(rc)_r = 0. \tag{2.15}$$

Evidently, Eq. (2.15) can be satisfied in three separate ways:

$$f(r) \equiv 0, \quad \omega_t(t) \equiv 0, \quad (rc)_r \equiv 0. \tag{2.16}$$

In the first two cases, Schwarzschild-like solutions are obtained, which, under certain conditions, give precisely the Schwarzschild solution. Case (iii) gives rise to situation where a new solution is obtained which leads to generalisations of the Eddington-Finkelstein transformation and the Kruskal–Szekeres coordinates.

2.5 Schwarzschild Solutions

It is immediately obvious, under the condition $f(r) = 0$, the off-diagonal metric tensor component is zero. Further, from Eqs. (2.9) and (2.12), we find

$$a(r,t) = \omega(t)(1 + r_0/r), \quad c(r,t) = r/(r + r_0), \tag{2.17}$$

where $\omega(t) = 1$ produces precisely the Schwarzschild solution.

The second case arises for $\omega_t(t) = 0$, or equivalently, $\omega = \alpha^2$ where α is a non-zero constant of integration (owing to the fact $\det g_{\alpha\beta} \neq 0$) and is chosen to be squared as a matter of convenience. By choosing $\omega(t) = \alpha^2$, we can conveniently express the solution in the form $ac + b^2 = \alpha^2$ where

$$\sqrt{ac} = \alpha \cos \psi_0, \quad b = \alpha \sin \psi_0, \tag{2.18}$$

for some constant angle ψ_0. Using the expressions derived for the functions a, b and c given by Eqs. (2.9), (2.13) and (2.12) respectively in conjunction with Eq. (2.18), we obtain an expression for $f(r)$, namely

$$f(r) = (r + r_0)\tan^2 \psi_0. \tag{2.19}$$

Finally, by substituting the above into Eq. (2.12) and (2.13) it can be shown

$$a(r,t) = \alpha^2(1 + r_0/r), \quad b(r,t) = \alpha \sin \psi_0, \quad c(r,t) = \frac{r}{r + r_0}\cos^2 \psi_0. \tag{2.20}$$

Again, the Schwarzschild solution is realised by imposing the conditions $\alpha = 1$ and $\psi_0 = 0$.

2.6 Derivation of New Solution

The third and final case to be considered is of particular interest as it produces a new spherically symmetric solution to the empty space field equations of general relativity. The remaining condition satisfying Eq. (2.15) is given by $(rc)_{,r} = 0$, where, upon integration yields $c(r) = \beta/r$, where β is a constant of integration. We obtain an expression for $f(r)$ by substituting $c(r) = \beta/r$ in Eq. (2.12) which is given by

$$f(r) = r^2/\beta - r - r_0, \tag{2.21}$$

and substituting (2.21) in (2.11) we find

$$b(r,t) = \left[\omega(t)\left(1 - \frac{\beta(r+r_0)}{r^2}\right)\right]^{1/2}. \tag{2.22}$$

The newly derived expressions for the functions a, b and c under the condition $(rc)_{,r}$ constitute an alternative exact solution of the field equations of general relativity where the line element is given by

$$ds^2 = \omega(t)\left(1 + \frac{r_0}{r}\right)dt^2 + 2\left[\omega(t)\left(1 - \frac{\beta(r+r_0)}{r^2}\right)\right]^{1/2} dr dt - \frac{\beta}{r}dr^2 - r^2 d\Omega^2. \tag{2.23}$$

Evidently, we may remove the arbitrary function of time $\omega(t)$ by introducing τ, such that $d\tau = \omega(t)^{1/2}dt$, in which case, Eq. (2.23) now reads

$$ds^2 = \left(1 + \frac{r_0}{r}\right)d\tau^2 + 2\left[1 - \frac{\beta(r+r_0)}{r^2}\right]^{1/2} dr d\tau - \frac{\beta}{r}dr^2 - r^2 d\Omega^2. \tag{2.24}$$

For $\beta = 0$ the line element becomes

$$ds^2 = \left(1 + \frac{r_0}{r}\right)d\tau^2 + 2 dr d\tau - r^2 d\Omega^2. \tag{2.25}$$

Using the Computer Algebra System (CAS) MAXIMA (Maxima 2016), the three line elements given by Eqs. (2.23), (2.24) and (2.25) can all be independently verified as bona fide solutions of the field equations, where α and β denote arbitrary constants and $\omega(t)$ denotes an arbitrary function of time.

2.7 Generalised Coordinates and Transformations

It is interesting to note that under the condition $\beta = 0$, the line element (2.25) corresponds to the well known outgoing Eddington–Finkelstein coordinates (Eddington

1924; Finkelstein 1958; Misner et al. 1973) which provide an analytic extension for the Schwarzschild radial coordinate from $r_0 < r < \infty$ to $0 < r < \infty$ (d'Inverno 1992). Since we have arbitrarily assigned the positive root for b in Eq. (2.22), then the corresponding ingoing Eddington–Finkelstein coordinates are obtained by adopting the negative root.

We further observe that both the Schwarzschild solution and Eq. (2.20) with $\alpha = 1$ have the common structure that can be written as

$$(\mathrm{d}s)^2 = \left(1 + \frac{r_0}{r}\right)\mathrm{d}\tau^2 + 2\sin\psi\ \mathrm{d}r\mathrm{d}\tau - \frac{\cos^2\psi}{1 + r_0/r}\mathrm{d}r^2 - r^2\mathrm{d}\Omega^2, \tag{2.26}$$

where ψ has the constant value ψ_0 for the Schwarzschild solution given by Eq. (2.20) and a variable value $\sin^{-1}([1 - \beta(r + r_0)/r^2]^{1/2})$ for the solution derived here involving the new constant β. As inferred by the so-called Birkhoff's theorem, the important question arises as to whether Eq. (2.24) is locally isometric to the Schwarzschild solution. The answer is in the affirmative and the specific details are given in Sect. 2.9. Although the new solution is locally isometric to the Schwarzschild solution, we note the new coordinate chart given by Eq. (2.24) does not embody the coordinate singularity at $r = r_0$ and allows for radially in/outgoing particles to pass freely between this region.

The line element given by Eq. (2.26) can be shown to become

$$(\mathrm{d}s)^2 = \left(1 + \frac{r_0}{r}\right)(\mathrm{d}\tau^2 + 2\sin\psi\ \mathrm{d}r^\star\mathrm{d}\tau - \cos^2\psi\ (\mathrm{d}r^\star)^2) - r^2\mathrm{d}\Omega^2, \tag{2.27}$$

where $\mathrm{d}r^\star$ and $\sin\psi$ are given by

$$\mathrm{d}r^\star = \frac{\mathrm{d}r}{1 + r_0/r}, \quad \sin\psi = \left[1 - \frac{\beta(r + r_0)}{r^2}\right]^{1/2}, \tag{2.28}$$

where $\psi = \pi/2$ corresponds to the case given by (2.25). By performing the transformation $\tau = \tau^\star + \rho\ (r^\star)$, it is clear that Eq. (2.27) becomes

$$\begin{aligned}(\mathrm{d}s)^2 = \left(1 + \frac{r_0}{r}\right)&\left\{\left[(\mathrm{d}\tau^\star)^2 + 2(\sin\psi + \rho\,'(r^\star))\mathrm{d}\tau^\star\mathrm{d}r^\star\right]\right.\\ &\left.+ \left[(\sin\psi + \rho\,'(r^\star))^2 - 1\right](\mathrm{d}r^\star)^2\right\} - r^2\mathrm{d}\Omega^2,\end{aligned} \tag{2.29}$$

where the prime denotes differentiation with respect to $r^\star$. From the structure of Eq. (2.29) it is apparent that $\rho\ (r^\star)$ can be chosen to produce any desired equation of the form of Eq. (2.27). Thus, as an example we may obtain the Schwarzschild line element by making

$$\frac{\mathrm{d}\rho}{\mathrm{d}r^\star} + \sin\psi = \sin\psi_0, \tag{2.30}$$

from which we may deduce that the function $\rho(r^\star)$ is determined by performing the integration

$$d\rho = \left\{ r \sin\psi_0 - \left[r^2 - \beta(r + r_0)\right]^{1/2} \right\} \frac{dr}{r + r_0}, \tag{2.31}$$

which clearly admits a range of analytical expressions depending upon the value of β, noting the greatly simplified form arising from the special case $\beta = -4r_0$.

As an illustration of the above, we derive the unknown function $\rho(r^\star)$ which allows for the coordinate transformation from Schwarzschild to Eddington–Finkelstein coordinates. In Schwarzschild geometry $\psi_0 = \beta = 0$ and Eq. (2.31) becomes

$$\rho(r) = \pm \int \frac{r}{r + r_0} dr = \pm \left[r - r_0 \ln(r + r_0) + C\right], \tag{2.32}$$

where C is a constant of integration. Substituting Eq. (2.32) into $\tau = \tau^\star + \rho(r)$ and applying to the Schwarzschild line element gives precisely the outgoing and ingoing Eddington–Finkelstein coordinates depending on the choice of sign. We note that $\tau = \tau^\star + \rho(r)$ together with Eq. (2.31) provides a generalisation of the Eddington–Finkelstein transformation.

Now, in order to extend the Kruskal–Szekeres coordinate transformation we define the variables ξ and η through the differential relations

$$d\xi = dr^\star + d\tau + \sin\psi\, dr^\star, \quad d\eta = dr^\star - d\tau - \sin\psi\, dr^\star, \tag{2.33}$$

where $r^\star$, τ and $\sin\psi$ are all as given above. Explicitly ξ and η are given by (derivation given in Appendix A.4)

$$\begin{aligned} \xi &= \tau + \int \frac{r}{r + r_0} \left\{ 1 + \left[1 - \frac{\beta(r + r_0)}{r^2}\right]^{1/2} \right\} dr, \\ &= \tau + r + \left[r^2 - \beta(r + r_0)\right]^{1/2} - (r_0 + \beta/2) \ln\left\{ \left[r^2 - \beta(r + r_0)\right]^{1/2} + r - \beta/2 \right\} \\ &\quad - r_0 \ln\left\{ \left[r^2 - \beta(r + r_0)\right]^{1/2} - r - \beta(r + r_0)/2r_0 \right\} + \xi_0, \end{aligned} \tag{2.34}$$

and

$$\begin{aligned} \eta &= -\tau + \int \frac{r}{r + r_0} \left\{ 1 - \left[1 - \frac{\beta(r + r_0)}{r^2}\right]^{1/2} \right\} dr, \\ &= -\tau + r - \left[r^2 - \beta(r + r_0)\right]^{1/2} + (r_0 + \beta/2) \ln\left\{ \left[r^2 - \beta(r + r_0)\right]^{1/2} + r - \beta/2 \right\} \\ &\quad - 2r_0 \ln(r + r_0) + r_0 \ln\left\{ \left[r^2 - \beta(r + r_0)\right]^{1/2} - r - \beta(r + r_0)/2r_0 \right\} + \eta_0, \end{aligned} \tag{2.35}$$

respectively, where ξ_0 and η_0 are introduced as arbitrary constants of integration, noting that ξ and η are essentially $r^\star + \tau^\star$ and $r^\star - \tau^\star$, and $\tau^\star$ is precisely as defined in the previous section.

In the above integral evaluations, it is assumed that the arguments of all logarithms are positive; in other cases, slightly modified formulae may apply. On evaluating the product $d\xi d\eta$ we see that Eq. (2.27) becomes

$$(ds)^2 = -\left(1 + \frac{r_0}{r}\right) d\xi d\eta - r^2 d\Omega^2, \tag{2.36}$$

where we can identify the product $d\xi d\eta$ as the generalised in/outgoing Eddington–Finkelstein coordinates.

Let us propose generalised coordinates $\mathcal{R}$ and $\mathcal{T}$ such that

$$\mathcal{R} + \mathcal{T} = e^{-\xi/2r_0}, \quad \mathcal{R} - \mathcal{T} = e^{-\eta/2r_0}, \tag{2.37}$$

as the new extended Kruskal–Szekeres coordinates for $\beta \neq 0$ where ξ and η are defined as in Eq. (2.33) since when $\psi = 0$ we have

$$\xi = \tau + r - r_0 \ln(r + r_0) + \xi_0, \quad \eta = -\tau + r - r_0 \ln(r + r_0) + \eta_0, \tag{2.38}$$

and on adopting the values $\xi_0 = \eta_0 = r_0 \ln(r_0)$ we may deduce the relations for the standard Kruskal–Szekeres coordinates

$$\begin{aligned} \mathcal{R} + \mathcal{T} &= \left(1 + \frac{r_0}{r}\right)^{1/2} e^{-r/2r_0} e^{-\tau/2r_0}, \\ \mathcal{R} - \mathcal{T} &= \left(1 + \frac{r_0}{r}\right)^{1/2} e^{-r/2r_0} e^{\tau/2r_0}, \end{aligned} \tag{2.39}$$

and therefore

$$\begin{aligned} \mathcal{R} &= \left(1 + \frac{r_0}{r}\right)^{1/2} e^{-r/2r_0} \cosh(\tau/2r_0), \\ \mathcal{T} &= -\left(1 + \frac{r_0}{r}\right)^{1/2} e^{-r/2r_0} \sinh(\tau/2r_0). \end{aligned} \tag{2.40}$$

where $\mathcal{R}$ and $\mathcal{T}$ denote standard Kruskal–Szekeres coordinates.

Finally, the line element given by Eq. (2.36) in generalised Kruskal–Szekeres coordinates is given by

$$(ds)^2 = 4\frac{r_0^3}{r} e^{r/r_0} (d\mathcal{T}^2 - d\mathcal{R}^2) - r^2 d\Omega^2. \tag{2.41}$$

We note that in general, the line element given by Eq. (2.41) applies for all β and is obvious from the relation

$$\xi + \eta = 2r - 2r_0 \ln\left(1 + \frac{r}{r_0}\right). \tag{2.42}$$

Thus, for all β we have with the above definition

$$\mathcal{R}^2 - \mathcal{T}^2 = \left(1 + \frac{r_0}{r}\right) e^{-r/r_0}, \tag{2.43}$$

assuming we adopt the same values for the arbitrary constants ξ_0 and η_0.

2.8 Equations of Motion

In this section, we derive the equations of motion, first integrals and Binet-like orbit equation (see d'Inverno (1992) for a discussion on Binet equations) associated with a test particle in the spacetime given by (2.24). The equations of motion can be used to approximately model the geodesic motion of near-Earth objects such as navigation satellites or solar system bodies. It is clear that the Lagrangian $\mathcal{L}$ and Hamiltonian $\mathcal{H}$ for a test body in the spacetime given by (2.24) have the same value and admit a first integral, which is a consequence of the fact that in general relativity there is no classical gravitational potential energy (Franklin 2010).

The Lagrangian associated with a test particle whose worldline is parameterised by the affine parameter s is given by

$$\begin{aligned}\mathcal{L} =& \frac{1}{2}\left\{ \left(1 + \frac{r_0}{r}\right)\left(\frac{\mathrm{d}\tau}{\mathrm{d}s}\right)^2 + 2\left[1 - \frac{\beta(r + r_0)}{r^2}\right]^{1/2}\left(\frac{\mathrm{d}r}{\mathrm{d}s}\right)\left(\frac{\mathrm{d}\tau}{\mathrm{d}s}\right)\right.\\ &\left. - \frac{\beta}{r}\left(\frac{\mathrm{d}r}{\mathrm{d}s}\right)^2 - r^2\left(\frac{\mathrm{d}\varphi}{\mathrm{d}s}\right)^2\right\},\end{aligned} \tag{2.44}$$

where, and without loss of generality, we have assumed the test particle is restricted to motion in the equatorial plane i.e. $\theta = \pi/2$. The non-zero Christoffel symbols of the second kind are given by

$$\Gamma^0_{00} = \frac{r_0\sqrt{r^2 - \beta(r + r_0)}}{2r^3}, \quad \Gamma^0_{11} = \frac{\beta\left(r^2 + \beta r_0\right)}{2r^3\sqrt{r^2 - \beta(r + r_0)}}, \quad \Gamma^0_{01} = -\frac{\beta r_0}{2r^3},$$

$$\Gamma^0_{22} = \sqrt{r^2 - \beta(r + r_0)}, \quad \Gamma^0_{33} = \sqrt{r^2 - \beta(r + r_0)}\sin^2\theta, \quad \Gamma^1_{00} = -\frac{r_0\left(r + r_0\right)}{2r^3},$$

$$\Gamma^1_{01} = -\Gamma^0_{00}, \quad \Gamma^1_{11} = -\Gamma^0_{01}, \quad \Gamma^1_{22} = -\left(r + r_0\right),$$

$$\Gamma^1_{33} = -\left(r + r_0\right)\sin^2\theta, \quad \Gamma^2_{12} = \frac{1}{r}, \quad \Gamma^2_{33} = -\cos\theta\sin\theta,$$

$$\Gamma^3_{13} = \Gamma^2_{12}, \quad \Gamma^3_{23} = \cot\theta,$$

and the second order geodesic equations of motion are given by

$$\frac{d^2\tau}{ds^2} = -\sqrt{r^2 - \beta(r + r_0)}\left[\frac{r_0}{2r^3}\left(\frac{d\tau}{ds}\right)^2 + \left(\frac{d\varphi}{ds}\right)^2\right]$$
$$- \frac{\beta\left(r^2 + \beta r_0\right)}{2r^3\sqrt{r^2 - \beta(r + r_0)}}\left(\frac{dr}{ds}\right)^2 + \frac{\beta r_0}{r^3}\left(\frac{d\tau}{ds}\right)\left(\frac{dr}{ds}\right), \tag{2.45}$$

$$\frac{d^2 r}{ds^2} = (r + r_0)\left[\frac{r_0}{2r^3}\left(\frac{d\tau}{ds}\right)^2 + \left(\frac{d\varphi}{ds}\right)^2\right]$$
$$- \frac{\beta r_0}{2r^3}\left(\frac{dr}{ds}\right)^2 + \frac{r_0\sqrt{r^2 - \beta(r + r_0)}}{r^3}\left(\frac{d\tau}{ds}\right)\left(\frac{dr}{ds}\right), \tag{2.46}$$

$$\frac{d^2\varphi}{ds^2} = -\frac{2}{r}\left(\frac{dr}{ds}\right)\left(\frac{d\varphi}{ds}\right). \tag{2.47}$$

We note the simplified form of (2.47) and the vanishing of the θ-geodesic equation are consequences of equatorial orbits.

Equation (2.47) is immediately integrable and yields a first integral corresponding to the conservation of angular momentum and is given by

$$r^2\frac{d\varphi}{ds} = \mathcal{J}, \tag{2.48}$$

which can be equivalently derived by observing φ does not appear explicitly in the Lagrangian (2.44). Hence, φ is known as a cyclic variable (Goldstein 1965) and corresponds to a conserved quantity. The third constant of the motion arises due to the time translation invariance of the Lagrangian and corresponds to the energy of the test particle $\mathcal{E}$ which is a consequence of Noether's theorem (Goldstein 1965) and is given by

$$\left(1 + \frac{r_0}{r}\right)\left(\frac{d\tau}{ds}\right) + \left[1 - \frac{\beta(r + r_0)}{r^2}\right]^{1/2}\left(\frac{dr}{ds}\right) = \mathcal{E}. \tag{2.49}$$

The Binet equation characterises the orbit of a test particle with position given as a function of the azimuth angle as opposed to time which is achieved through the coordinate transformation $r = u^{-1}$ where $u = u(\varphi)$. Substituting Eq. (2.49) in (2.44) we find

$$1 = \left(1 + \frac{r_0}{r}\right)^{-1}\left\{\mathcal{E}^2 - 2\mathcal{E}\frac{dr}{ds}\left[1 - \frac{\beta(r + r_0)}{r^2}\right]^{1/2} + \left[1 - \frac{\beta(r + r_0)}{r^2}\right]\left(\frac{dr}{ds}\right)^2\right\}$$
$$+ 2\left(\frac{dr}{ds}\right)\left(1 - \frac{\beta(r + r_0)}{r^2}\right)^{1/2}\left[\mathcal{E} - \left(\frac{dr}{ds}\right)\left(1 + \frac{r_0}{r}\right)^{-1}\right] - \frac{\beta}{r}\left(\frac{dr}{ds}\right) - r^2\left(\frac{d\varphi}{ds}\right), \tag{2.50}$$

where we have used $\mathcal{L} \equiv 1/2$. The above equation is simplified by exploiting the conservation of angular momentum and is given by

$$1 = \left(1 + \frac{r_0}{r}\right)^{-1} \left\{ \mathcal{E}^2 - \left[\frac{\mathcal{J}}{r^2}\left(\frac{\mathrm{d}r}{\mathrm{d}s}\right)\right]^2 \left(1 - \frac{\beta(r + r_0)}{r^2}\right)\right\} - \frac{\mathcal{J}\beta}{r^5}\left(\frac{\mathrm{d}r}{\mathrm{d}s}\right) - \frac{\mathcal{J}^2}{r^2}, \tag{2.51}$$

which, after further simplification and introducing the Binet transformation yields

$$\frac{\mathrm{d}^2 u}{\mathrm{d}\varphi^2} + u = \frac{GM}{\mathcal{J}^2} + 3GMu^2. \tag{2.52}$$

We note the approximate solution to Eq. (2.52) is obtained via perturbation methods about the linear harmonic oscillator and can be found in the undergraduate literature (see for example Dalarsson and Dalarsson 2015; d'Inverno 1992). The exact analytical solution is obtained in terms of elliptic functions and has been explicitly derived by many authors (see Hagihara 1930; Darwin 1959; Rodrŕguez 1987).

2.9 Conclusions

We have derived a spherically symmetric, static solution of the empty space field equations of general relativity, where the line element given by Eq. (2.24) involves two arbitrary constants. In the weak field limit (d'Inverno 1992), we identify $r_0 = -2GM/c^2$ as the Schwarzschild radius. The second arbitrary constant allows for the generalisations of the well-known Eddington–Finkelstein transformation and the Kruskal–Szekeres coordinates. Furthermore, a variety of analytical forms for the extended Eddington–Finkelstein transformation and the Kruskal–Szekeres coordinates are derivable depending on the value of β, and one derivation is provided in Appendix A.4.

The key point is that this solution does not arise if the usual assumptions leading to the Schwarzschild solution are made at the outset. We note the Schwarzschild solution is formally obtained by introducing a new time variable τ^* in Eq. (2.24) through the differential relation given by

$$\mathrm{d}\tau^* = \mathrm{d}\tau + \frac{[1 - \beta(r + r_0)/r^2]^{1/2}}{1 + r_0/r}\mathrm{d}r. \tag{2.53}$$

The Binet equation (2.52) given in Sect. 2.8 is equivalent to the Binet equation for a test particle in Schwarzschild geometry. It is not immediately obvious that this would be the resulting formalism as the coordinate charts, Christoffel symbols and geodesic equations for the spacetime (2.24) and the Schwarzschild solution are starkly different. Using standard perturbation techniques found in the undergraduate literature for general relativity, it can be shown that the approximate solution to (2.52) accurately describes the relativistic contribution to planetary perihelion precession.

As a final remark, let us introduce a new time coordinate w in the Schwarzschild line element (2.1) through the differential relation

$$\mathrm{d}w = \mathrm{d}t + \frac{\mathrm{d}f}{\mathrm{d}r}\mathrm{d}r,$$

where $f = f(r)$ is *any* arbitrary, radially dependent function, then the resulting line element $\mathrm{d}s^*$ is given by

$$\begin{aligned}(\mathrm{d}s^*)^2 &= \left(1+\frac{r_0}{r}\right)\mathrm{d}w^2 - 2\left(1+\frac{r_0}{r}\right)\frac{\mathrm{d}f}{\mathrm{d}r}\mathrm{d}r\mathrm{d}w \\ &\quad - \left(\left(1+\frac{r_0}{r}\right)^{-1} - \left(1+\frac{r_0}{r}\right)\left(\frac{\mathrm{d}f}{\mathrm{d}r}\right)^2\right)\mathrm{d}r^2 - r^2\mathrm{d}\Omega^2.\end{aligned} \tag{2.54}$$

It has been observed using the in-built algebraic tensor packages `ctensor` and `itensor` of Maxima (2016), the line element (2.54) containing *any* arbitrary radially dependent function is necessarily a solution to the field equations of general relativity. The fact that the line element (2.54) satisfies the Einstein field equations implies the existence of an infinite number of spherically symmetric solutions which are equivalent to the Schwarzschild solution (2.1) up to a coordinate transformation. This property of the Birkhoff theorem was discussed in Florides and Synge (1961) who derive a series of algebraic and differential conditions for the metric components which can be used to determine whether a given line element is simply the Schwarzschild solution "in disguise" (Florides and Synge 1961).

Part III
The Post-Newtonian Formalism

Chapter 3
The Post-Newtonian Approximation

3.1 Nomenclature

ϵ	is the post-Newtonian dimensionless expansion parameter
v	is the velocity of a material body
c	is the speed of light in vacuum
U	is the Newtonian gravitational potential
G	is Newton's gravitational constant
M, R	are the mass and radius of a central gravitating body
$\gamma, \beta, \alpha_i, \zeta_\sigma$	are the post-Newtonian parameters
$g_{\alpha\beta}, G_{\alpha\beta}$	are the global and local metric tensor components respectively
x^α	are the global coordinates
X_A^α	are the local coordinates for a body A
$\mathfrak{g}^{\mu\nu}$	are the components of the gothic metric tensor
g	is the determinant of the metric tensor
$R_{\mu\nu}$	are the components of the Ricci tensor
R	is the Ricci scalar
γ_{ij}	are the components of a flat three-metric
δ_{ij}	is the Kronecker delta function
$T_{\mu\nu}$	are the components of the energy momentum tensor
σ, σ^i	are the global active mass and mass-current densities.

J. O'Leary, *General Relativistic and Post-Newtonian Dynamics for Near-Earth Objects and Solar System Bodies*, Springer Theses,
https://doi.org/10.1007/978-3-030-80185-4_3

3.2 Introduction

The motion of near-Earth objects and solar system bodies is primarily dictated by gravity.[1] Hence, the accurate modelling of gravitational effects in the solar system holds the key to increasing accuracy of present and future space-geodetic techniques (Müller et al. 2008), interplanetary navigation (Moyer 1971, 2005) and global navigation satellite systems (Ashby 2003). Due to the slow motion, weak field nature of solar system bodies, it is currently sufficient to account for general relativity using approximation methods. The post-Newtonian formalism is a method developed to approximately solve the Einstein field equations by formally expanding the metric tensor components using the dimensionless parameter $\epsilon = v/c$ where the velocity of a given body and the speed of light in vacuum are given by v and c respectively. We may assign limits for the velocity and gravitational potential U of any solar system body according to

$$\frac{U}{c^2} \sim \left(\frac{v}{c}\right)^2 < 10^{-5}, \tag{3.1}$$

so that for example, we have the following upper limits for the Earth (E) (Soffel 2014, Chap. 4)

$$\frac{U_E}{c^2} \simeq 6.9 \times 10^{-10}, \qquad \left(\frac{v_E}{c}\right)^2 \simeq 10^{-8}, \tag{3.2}$$

where $U_E \equiv GM_E/R_E$ and the mass and radius of the Earth are given by M_E and R_E respectively.

The complete treatment of general relativistic celestial mechanics is given by the Damour, Soffel, Xu formalism (see Sect. 3.4) which provides the mathematical framework required to describe the interaction of n arbitrarily shaped, rotating and deformable bodies at the first post-Newtonian level. The present chapter is intended to serve as a self-contained introduction to the early historical developments of the post-Newtonian approximation and discuss the mathematical development of the modern Damour, Soffel, Xu framework which will be the main topic for Part III of the thesis.

3.3 Background

Shortly after the publication of Einstein's general theory of relativity, approximation methods were developed to combat the difficulties associated with the non-linear structure of the field equations and derive equations of motion associated with realistic astrophysical systems. Assuming bodies are slowly moving and produce weak gravitational fields, the post-Newtonian approximation has been successfully

[1] As noted, gravity is the dominant perturbation in orbit determination. However, the modelling of atmospheric drag in the low Earth orbit region is the most uncertain and is regarded as the primary limitation for reliable orbit determination.

developed over the past century and aptly coined *unreasonably effective* (Will 2011) in the description of gravitational physics. Early developments of the post-Newtonian approximation are credited to Droste (1917); De Sitter (1916); Eddington and Clark (1938); Lorentz and Droste (1937); Einstein et al. (1938) who developed a primitive post-Newtonian framework which provided an ample opportunity to:

- Compare general relativity with classical Newtonian gravity and identify observable relativistic phenomenon such as planetary perihelion precession. We note Will (2018) has identified a new relativistic contribution to the perihelion advance of Mercury which is greater than the second-order post-Newtonian contribution by a factor of 100. The new contribution arises in the post-Newtonian equations of motion as a consequence of the interaction between Mercury and the other solar system bodies. Further, Tucker and Will (2019) provide a detailed review of standard methods adopted in the literature used to calculate perihelion precession for high-order post-Newtonian approximations.
- Derive the general relativistic n-body equations of motion formally known as the *Einstein–Infeld–Hoffmann* equations (Einstein et al. 1938). Derived explicitly in Chap. 5, the Einstein–Infeld–Hoffmann equations form the basis for the American, Russian and French state-of-the-art solar system ephemerides, namely, the DE (Folkner et al. 2014), EPM[2] and INPOP (Viswanathan et al. 2018 series respectively. Each ephemeris numerically integrates the Einstein–Infeld–Hoffmann equations for at least 300 independent solar system bodies which accounts for the major planets, moons and asteroids. More recently, in Suzuki et al. (2019), the authors numerically integrate the Einstein–Infeld–Hoffmann equations for a hierarchical three body system (inner binary pulsar system with a tertiary outer perturbing body), and determine the time evolution of the associated orbital elements and calculate the corresponding shift of periastron time of the inner binary system.

Further developments to the post-Newtonian approximation are given by:

- Fock (1964) and Chandrasekhar (1965) independently advanced the framework of the early first post-Newtonian approximation of gravitationally interacting point particles to the modern, hydrodynamical formalism using the post-Newtonian extension of the equations of fluid dynamics. Further, Chandrasekhar (1969) and Chandrasekhar and Nutku (1969) extended the hydrodynamical description of arbitrarily shaped extended bodies to higher orders of approximation and provided the corresponding first integrals.
- Nordtvedt (1968); Thorne and Will (1971); Will (1971a, b) and Will and Nordtvedt (1972) systematically develop a theoretical framework to compare predictions of general relativity with experiment. This was achieved with the introduction of a series of parameters $(\gamma, \beta, \alpha_i, \zeta_\sigma)$ as the coefficients of terms in the post-Newtonian metric and is known as the parameterised post-Newtonian formalism. The parameters are as follows:

[2]See http://iaaras.ru/en/dept/ephemeris/epm/2017/ for up-to-date details.

1. Spacetime curvature – γ,
2. Nonlinearity of gravity – β,
3. Preferred global rest-frame – α_i,
4. Conservation laws for energy-momentum – ζ_σ.

We note the parameters in general relativity have values $\alpha_i = \zeta_\sigma = 0$ and $\gamma = \beta = 1$ respectively. Eddington (1920) introduced the parameters γ, β in the seminal treatise *the mathematical theory of relativity* (see Chap. 3 Sect. 47) and first discusses methods to test general relativity with experiment.

The preceding discussion provides an overview of the early historical developments of the post-Newtonian formalism, and the reader is referred to the following review articles for complete discussions on:

- The problem of motion in classical Newtonian theory and general relativity (Damour 1987).
- The effectiveness of the post-Newtonian formalism at higher orders with applications to stronger field regimes. Examples include the motion of binary pulsar systems (Will 2011).
- Experimental tests of gravitational theories using the parameterised post-Newtonian framework (Ashtekar 2005, Chap. 8).
- An overview of the derivation of the post-Newtonian hydrodynamical equations of motion and the corresponding parameterised post-Newtonian equations (Peron et al. 2015, Chap. 1).
- Modern relativistic celestial mechanics and the corresponding mathematical framework (Battista et al. 2017).

3.4 The DSX Formalism

In the concluding paragraphs of the influential review article, *The problem of motion in Newtonian and Einsteinian gravity*, Damour (1987) addresses some pertinent defects within the field of *relativistic celestial mechanics* at that particular time. Issues ranged from repetition of known results to ambiguities and typographical errors within the post-Newtonian literature. Other issues raised by Damour (1987) included the development of hybrid post-Newtonian theories which led to confusion, as well as further issues related to inconsistent notation amongst the community. A major breakthrough for the field of relativistic celestial mechanics came with the publication of a series of papers by Damour, Soffel, Xu (hereafter DSX) and is given in Damour et al. (1991, 1992, 1993, 1994). The DSX formalism is a consistent theory of relativistic celestial mechanics complete at the first post-Newtonian level. The DSX framework lays the foundation in the definition of modern relativistic reference frames and will be the main focus of the remainder of this chapter which is based on the following bodies of work: Soffel (2014, Chap. 4), Damour (1987); Damour et al. (1991, 1992) and more recently Battista et al. (2017).

The motivation behind the development of the DSX formalism stems from a series of technical and conceptual issues associated with the *original* post-Newtonian approximation which are discussed in great detail in Soffel (2014, Chap. 4), Damour (1987); Damour et al. (1991, 1992). The major issues can be summarised as:

- The relativistic $n-$body problem was originally modelled using a single, global coordinate chart approach. Accordingly, a body which is spherically symmetric with respect to its rest frame, will appear to be deformed in the global frame due to its orbital velocity, formally known as Lorentz contraction.
- Relativistic celestial mechanics was originally modelled as close as possible to classical celestial mechanics. This led to the introduction and misuse of a Newtonian-like, or, what is more commonly referred to in the literature, a "neo-Newtonian" (Damour 1987; Soffel 2014) approach to the post-Newtonian approximation, where the global temporal and spatial coordinates become inherently associated with the absolute space and time of classical Newtonian theory. Similarly, this gave rise to confusion if the aforementioned *relativistic deformations* were not accounted for.

In contrast to this, the DSX formalism considers the problem of describing n gravitationally interacting bodies using $n + 1$ coordinate charts which are decomposed as follows. A single, global reference frame describing the overall dynamics of the n bodies and a further n local reference frames describing the individual structure of each body (see Fig. 3.1). Traditionally, the notation adopted in the DSX formalism is that quantities in local coordinate systems are denoted using uppercase symbols and

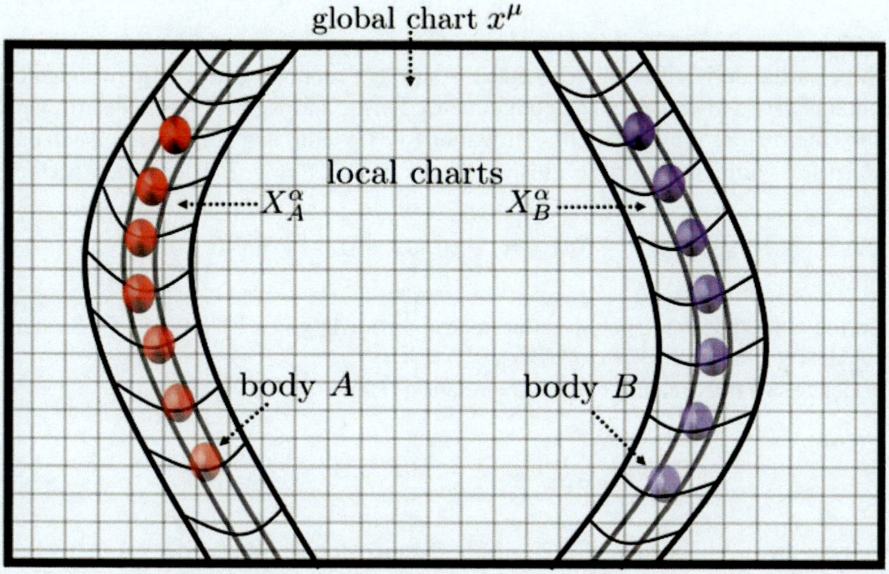

Fig. 3.1 $n + 1$ coordinate chart approach to the n-body problem in the DSX formalism

quantities in the global frame are denoted using lowercase symbols. For example, the metric tensor components in local and global frames are denoted $G_{\alpha\beta}$ and $g_{\alpha\beta}$ respectively.

The DSX formalism takes advantage of an exponential parameterisation of the metric tensor for each of the $n+1$ coordinate charts (see Blanchet and Damour (1989)) which formally linearises the resulting field equations. Further, the exponential parameterisation relies only on two potentials: a single scalar *gravitoelectric* potential and a vector *gravitomagnetic* potential. Hence, such a parameterisation is more compact than the standard post-Newtonian counterparts which often contains additional complicated relativistic potentials.

3.5 Algebraic Gauge Conditions

The standard approach in post-Newtonian theory is to introduce differential gauge conditions such as the harmonic or standard post-Newtonian gauge which aid in simplifying the Einstein field equations. In contrast to this, Damour et al. (1991) employ algebraic conditions for the metric tensor components, leaving a freedom in the choice of temporal coordinate; while completely fixing the spatial components.

Let us follow directly the approach outlined in Damour et al. (1991) and introduce the contravariant *gothic metric* $\mathfrak{g}^{\mu\nu} \equiv g^{\mu\nu}\sqrt{-\det g_{\mu\nu}}$ (see for example Landau (2013)) in the Einstein tensor

$$2\,g\left[R^{\mu\nu} - \frac{1}{2}Rg^{\mu\nu}\right] \equiv \partial_{\rho\sigma}\left(\mathfrak{g}^{\mu\nu}\mathfrak{g}^{\rho\sigma} - \mathfrak{g}^{\mu\rho}\mathfrak{g}^{\nu\sigma}\right) + Q^{\mu\nu}, \tag{3.3}$$

where partial derivatives are denoted by ∂_ρ and $Q^{\mu\nu}$ denotes a quadratic form which contains first derivatives of the gothic metric tensor and we denote the determinant of the metric tensor by $g = \det g_{\mu\nu}$. Equation (3.3) is valid in all dimensions such that the spatial components of the Einstein tensor ($i, j, k, l = 1, 2, 3$) can be expressed as

$$2\,g\left[R^{ij} - \frac{1}{2}Rg^{ij}\right] \equiv \partial_{kl}\left(\mathfrak{g}^{ij}\mathfrak{g}^{kl} - \mathfrak{g}^{ik}\mathfrak{g}^{jl}\right) + O(c^{-4}), \tag{3.4}$$

and the corresponding spatial gothic metric tensor is given by $\mathfrak{g}^{ij} = \sqrt{\gamma}\gamma^{ij}$ where $\gamma \equiv \det\left(\gamma_{ij}\right)$ and $\gamma^{ij} = -g_{00}g_{ij} + g_{0i}g_{0j}$.

Hence, we may readily deduce the relation (Damour et al. 1991)

$$g\left[R^{ij}(g) - \frac{1}{2}R(g)g^{ij}\right] = \gamma\left[R^{ij}(\gamma) - \frac{1}{2}R(\gamma)g^{ij}\right]. \tag{3.5}$$

We note the left-hand side of Eq. (3.5) is 4-dimensional and of order $O(c^{-4})$ while the right-hand side is three-dimensional. Damour et al. (1991) then argue that if $g_{\mu\nu}$ is a post-Newtonian solution to the Einstein field equations, then the corresponding three-metric γ_{ij} has Ricci and Riemann tensors of order $O(c^{-4})$. In other words, the three-metric is flat and satisfies the spatial Einstein tensor to appropriate order. This suggests the existence of a preferred class of spatial coordinates which are conformally Cartesian which are subject to the relations

$$\gamma_{ij} = \delta_{ij} + O(c^{-4}), \qquad \mathfrak{g}^{ij} = \delta^{ij} + O(c^{-4}), \qquad -g_{00}g_{ij} = \delta_{ij} + O(c^{-4}). \quad (3.6)$$

The above expressions then fix the spatial coordinates up to a time dependent coordinate transformation (Damour et al. 1991).

3.6 Post-Newtonian Field Equations

The standard post-Newtonian assumptions are as follows: the metric tensor components deviate from flat Minkowskian space by some small perturbation according to

$$g_{00} = -1 + O(c^{-2}), \qquad g_{0i} = O(c^{-3}), \qquad g_{ij} = \delta_{ij} + O(c^{-2}), \quad (3.7)$$

and the components of the energy-momentum tensor are of order $O(c^2, c^1, c^0)$. According to Blanchet and Damour (1989), the components of $g_{\mu\nu}$ are given by

$$\begin{aligned} g_{00} &= -\exp\left(-\frac{2w}{c^2}\right) \simeq -1 + \frac{2w}{c^2} - \frac{2w^2}{c^4} + O(c^{-6}), \\ g_{0i} &= -\frac{4w_i}{c^3} + O(c^{-5}), \\ g_{ij} &= \gamma_{ij}\exp\left(\frac{2w}{c^2}\right) \simeq \gamma_{ij}\left(1 + \frac{2w}{c^2}\right) + O(c^{-4}), \end{aligned} \quad (3.8)$$

where the γ_{ij} are defined in Sect. 3.5, γ^{ij} denotes the inverse of γ_{ij} and $w^i \equiv \gamma^{ij} w_j$. The metric tensor components $g_{\mu\nu}$ then consist of a scalar potential w, a vector potential w^i and a symmetric second rank tensor γ_{ij} with one, three and six degrees of freedom respectively.

Now since γ_{ij} satisfies the spatial Einstein field equations to post-Newtonian order, the remaining four independent Einstein equations are given by (Damour 1987; Damour et al. 1991)

$$R^{00} = \frac{8\pi G}{c^4}\left(T^{00} - \frac{1}{2}T g^{00}\right), \qquad R^{0i} = \frac{8\pi G}{c^4}\left(T^{0i} - \frac{1}{2}T g^{0i}\right), \quad (3.9)$$

where R^{00} and R^{0i} are given by

$$\begin{aligned} R^{00} &\equiv -\frac{1}{c^2}\nabla^2 w - \frac{1}{c^4}\left[3\partial_{tt} w + 4\partial_{ti} w_i\right] + O(c^{-6}), \\ R^{0i} &\equiv -\frac{2}{c^3}\left[\nabla^2 w_i - \partial_{ij} w_j - \partial_{ti} w\right] + O(c^{-5}), \end{aligned} \tag{3.10}$$

and we note Laplace's differential operator is given by ∇^2. The remaining Einstein equations take the form

$$\begin{aligned} \nabla^2 w + \frac{3}{c^2}\partial_{tt} w + \frac{4}{c^2}\partial_{ti} w_i &= -4\pi G\sigma, \\ \nabla^2 w_i - \partial_{ij} w_j - \partial_{ti} w &= -4\pi G\sigma^i, \end{aligned} \tag{3.11}$$

where we have introduced the quantities σ and σ^i according to

$$\sigma \equiv \frac{\left(T^{00} + T^{ss}\right)}{c^2}, \qquad \sigma^i \equiv \frac{T^{0i}}{c}. \tag{3.12}$$

Blanchet and Damour (1989); Damour et al. (1991) define σ and σ^i as an active gravitational mass density and an active mass current density respectively. According to DSX, σ generalises the density ρ of Newtonian theory where σ^i does not have a Newtonian analogue in terms of classical field generating sources (Soffel 2014, Chap. 4).

Evidently, the conformally Cartesian coordinates are harmonic such that

$$\Box_g x^i = 0 + O(c^{-4}), \tag{3.13}$$

where the Laplace-Beltrami differential operator (in 4-dimensional psuedo-Riemannian space) is denoted by $\Box$ and is given by

$$\Box_g f(x^\mu) \equiv \frac{1}{\sqrt{g}}\partial_\mu\left(\sqrt{g}g^{\mu\nu}\partial_\nu f\right), \tag{3.14}$$

and the only remaining freedom is the choice of temporal coordinate x^0. By imposing the condition that x^0 is harmonic such that

$$\Box_g x^0 = -\frac{4}{c^3}\left(\partial_t w + \partial_i w_i\right) = 0, \tag{3.15}$$

then the field equations are given in final simplified linear form by

$$\Box w = -4\pi G\sigma, \qquad \nabla^2 w_i = -4\pi G\sigma^i. \tag{3.16}$$

It is important to note that in the outline above, attention has been restricted to the global reference frame. Similar expressions exist for the field equations in the local frame and will be the topic of interest in the following chapter.

Chapter 4
Post-Newtonian Energy Integrals

4.1 Overview

In this chapter, we consider the first integral for energy associated with the post-Newtonian Kepler problem. The original contributions of this chapter are:

- the derivation of a new energy integral associated with the first post-Newtonian equations of motion in the local geocentric reference frame.
- Employing high-order symplectic numerical integration schemes to compare the newly obtained energy integral with the corresponding quantities in the literature.

This chapter is organised as follows: in Sect. 4.4 we give an overview of the independent Einstein field equations and derive the equations of motion associated with a near-Earth object in the local geocentric reference frame. In Sect. 4.5 we derive a Jacobi-like energy integral which is obtained by algebraic manipulation of the equations of motion associated with the post-Newtonian regime. In Sect. 4.6 we give an independent derivation of the new energy integral from a variational principle which arises due to the time translation symmetry of the stationary gravitational field. Finally, in Sects. 4.7 and 4.8 we present numerical results which:

- provide a further independent verification of the new energy integral by numerically integrating the full first post-Newtonian equations of motion.
- Compare the error associated with energy integrals found in the literature with the first integral derived in Sect. 4.5. This is achieved by exploiting the structure-preserving properties of high-order symplectic numerical integration schemes.

The following chapter and the results discussed have been published in O'Leary et al. (2018) and O'Leary and Hill (2018).

- O'Leary, J., Hill, J.M. and Bennett, J.C., 2018. On the energy integral for first post-Newtonian approximation. *Celestial Mechanics and Dynamical Astronomy*, 130(7), p. 44.

J. O'Leary, *General Relativistic and Post-Newtonian Dynamics for Near-Earth Objects and Solar System Bodies*, Springer Theses,
https://doi.org/10.1007/978-3-030-80185-4_4

- O'Leary, J. and Hill, J.M., 2018. Numerical conservation of exact and approximate first post-Newtonian energy integrals. In: *The Advanced Maui Optical and Space Surveillance Technologies Conference.*

4.2 Nomenclature

$G_{\mu\nu}$	are the local metric tensor components
W, W^j	are the generalised local gravitational scalar and vector potentials
c	is the speed of light in vacuum
Σ, Σ^j	are the local active gravitational mass and mass-current densities
X^j	are the local geocentric Cartesian coordinates
T	is the geocentric coordinate time
V^α	are the components of the local coordinate velocity
$\mathcal{L}$	is the first post-Newtonian Lagrangian function
$\mathcal{E}$	is the test particle energy function per unit mass
r	is the magnitude of the local geocentric coordinates
ϵ_{ijk}	is the fully antisymmetric Levi-Civita symbol
G	is Newton's gravitational constant
S^i	is the angular momentum vector of the Earth.

4.3 Background

The general theory of relativity is now widely accepted as providing the most accurate theory of gravity. However, the field equations of general relativity (Griffiths and Podolský 2009; Stephani et al. 2009) are a system of ten non-linear, coupled partial differential equations, where known solutions often exhibit such high degrees of symmetry that modelling realistic astrophysical situations is often restricted (Will 2011).

To model physically realisable situations, approximation methods have been developed to determine n-body equations of motion (see Chap. 5) and approximate solutions of the field equations of gravitation (see Chaps. 2 and 3). The post-Newtonian approximation was originally established to describe the spacetime associated with weak gravitational fields produced by slowly moving bodies e.g. the planets in our solar system. However, the post-Newtonian formalism is not restricted to the weak-field regime as expected and when approximated to a suitably high order can be used in the study of strong field phenomena such as the coalescence of black hole binary systems (Poisson and Will 2014).

The post-Newtonian regime is widely exploited in both geodesy and astronomy, and through a series of resolutions (1991–2000) (Soffel et al. 2003), the International Astronomical Union (IAU) now recommends that general relativity be used as the framework in current definitions of reference systems. Currently, the IAU recommend adopting the DSX framework (see Chap. 3) using a global solar system barycentric frame and a local geocentric frame. Typical uses for solar system barycentric and local geocentric frames are the modelling of light propagation from distant solar system bodies and the motion of near-Earth objects such as navigation satellites (Soffel et al. 2003). The IAU-recommended metric tensor components are given by Eqs. (3.8) or (4.1) depending on the problem. It should be noted that the IAU suggest the addition of the non-linear potential term arising in the time-time component of the metric tensor (g_{00} or G_{00} respectively) is necessary for the derivation of the Einstein–Infeld–Hoffmann equations. However, as observed in Brumberg (2007) and Brumberg (2017), the inclusion of the non-linear potential term is only required when deriving the equations of motion from a geodesic principle. In fact, the Einstein–Infeld–Hoffmann equations are derivable from the linearised metric of general relativity which is demonstrated explicitly in Brumberg (2007).

The IAU established the International Earth Rotation and Reference System Service (IERS) for celestial and terrestrial reference systems providing accurate and consistent connecting transformations. The IERS implemented the resolutions of the IAU and now consider celestial and terrestrial reference systems as four-dimensional reference systems connected by relativistic transformations (Brumberg and Groten 2001). Further, the IERS maintain and publish the post-Newtonian equations of motion which need be included as effective forces in orbit determination software. The post-Newtonian perturbations account for Schwarzschild, Lense-Thirring and DeSitter terms and are discussed in Petit and Luzum (2010).

Hugentobler (2010) employs Gaussian planetary perturbation equations and determines the magnitude each individual relativistic term has on the orbital elements. The author demonstrates the magnitude of each effect for several test cases including GPS satellites and both CHAMP[1] and LAGEOS[2] geodetic missions. It is the opinion of the present author, that the IERS-recommended post-Newtonian equations of motion are restrictive, and truncated to include only a monopole term.

Brumberg (2004) derives intermediary equations of motion which consider some additional post-Newtonian effects, such as the post-Newtonian quadrupole moment of the Earth. These equations provide an improved alternative to those recommended by the IERS; while avoiding the explicit and complicated full post-Newtonian equations of motion (see for example Damour et al. 1994; Brumberg 2017).

Roh et al. (2016) developed the high-fidelity orbit propagation software KASIOP (Roh and Choi 2014) and suggests, at least for the LARES misison,[3] the post-Newtonian quadrupole moment of the Earth should be considered in future orbit determination. Further, Roh et al. (2016) demonstrated that the effect of the post-Newtonian quadrupole moment of the Earth is larger in magnitude than the contribution from atmospheric drag. It should be noted; this is a consequence of the high

[1]See Pail (2014) for an overview of Earth gravity field determination missions.

[2]See Soffel and Langhans (2012, Chap. 8) for an overview of LAGEOS and satellite laser ranging in general.

[3]See Ciufolini et al. (2012) for an overview of the LARES mission.

mass-to-surface area of the LARES mission and so is an expected result (Ciufolini et al. 2012).

When considering post-Newtonian effects, other than those recommended by the IERS, the exhaustive study by Roh et al. (2016) suggests care should be taken to avoid modelling redundant perturbations or failing to include those which are relevant. Factors such as orbital regime, mass-to-surface area and other engineering factors should be carefully considered.

4.4 Field Equations and Equations of Motion

In this section, we provide the necessary background to derive the equations of motion associated with a near-Earth satellite in the local geocentric reference frame. Unless stated otherwise, the conventions adopted in this chapter are: Latin indices run over spatial variables $i, j = 1, 2, 3$ while Greek indices account for spacetime coordinates $\mu, \nu = 0, 1, 2, 3$ with the 0th component being reserved for time and quantities in the local geocentric frame are denoted using uppercase letters (see Chap. 3).

In general relativity, the gravitational field is described by ten independent relativistic potentials given by the metric tensor $G_{\mu\nu}$. The local metric tensor components $G_{\mu\nu}$ are prescribed according to (Blanchet and Damour 1989; Damour et al. 1991)

$$
\begin{aligned}
G_{00} &= -\exp\left(-\frac{2W}{c^2}\right) \simeq -1 + \frac{2W}{c^2} - \frac{2W^2}{c^4} + O(c^{-6}), \\
G_{0j} &= -\frac{4W_j}{c^3} + O(c^{-5}), \\
G_{ij} &= \delta_{ij}\exp\left(\frac{2W}{c^2}\right) \simeq \delta_{ij}\left(1 + \frac{2W}{c^2}\right) + O(c^{-4}),
\end{aligned}
\tag{4.1}
$$

where the scalar and vector gravitational potentials, speed of light in vacuum and the usual Kronecker delta function are given by W, W^j, c and δ_{jk} respectively. We have implicitly assumed the coordinate conditions (3.6) are also satisfied in the local geocentric frame. Hence, using the same formal arguments from Chap. 3, the remaining non-zero Einstein field equations satisfied by W and W^j in the local frame are given by (Damour et al. 1991)

$$
\begin{aligned}
\Box W + \frac{4}{c^2}\partial_T\left(\partial_T W + \partial_j W_j\right) &= -4\pi G\Sigma, \\
\nabla^2 W_j - \partial_j\left(\partial_T W + \partial_k W_k\right) &= -4\pi G\Sigma^j.
\end{aligned}
\tag{4.2}
$$

We note the quantities Σ and Σ^j are the local geocentric generalisations of the global source terms given by Eqs. (3.12).

For the remainder of this chapter, we assume the gravitational potentials are stationary, such that, the scalar and vector potentials are assumed to be spatially dependant only. For example, $W \equiv W(X^j)$ where $X^j \equiv (X, Y, Z)$ are the usual geocentric Cartesian coordinates associated with the relativistic geocentric celestial reference system (Soffel and Langhans 2012).

The non-zero, independent Christoffel symbols of the second kind associated with (4.1) are given by (Poisson and Will 2014)

$$
\begin{aligned}
\Gamma^0_{00} &= -\frac{1}{c^3}\partial_T W, \\
\Gamma^j_{00} &= -\frac{1}{c^2}\left(\partial_j W - \frac{1}{c^2}4W\partial_j W + \frac{4}{c^2}\partial_T W_j\right), \\
\Gamma^j_{0k} &= \frac{1}{c^3}\delta_{jk}\partial_T W - \frac{2}{c^3}\left(\partial_k W^j - \partial_j W^k\right), \\
\Gamma^j_{ik} &= \frac{1}{c^2}\left(\delta_{jk}\partial_i W + \delta_{ij}\partial_k W - \delta_{ik}\partial_j W\right), \\
\Gamma^0_{0j} &= -\frac{1}{c^2}\partial_j W, \\
\Gamma^0_{ik} &= \frac{1}{c^3}\delta_{ik}\partial_T W + \frac{2}{c^3}\left(\partial_i W^k + \partial_k W^i\right),
\end{aligned}
\tag{4.3}
$$

and we have retained the full set of Christoffel symbols here for spatially *and* temporarily dependent scalar and vector potentials which will be required in order to derive the Einstein–Infeld–Hoffmann equations in Chap. 5.

For near-Earth objects such as geodetic satellites, the worldline of a satellite is not parameterised by its proper time τ (as is often the case in general relativity) but rather by the geocentric coordinate time (TCG)[4] and therefore, the usual geodesic equation of motion

$$
\frac{d^2X^\alpha}{d\tau^2} = -\Gamma^\alpha_{\beta\gamma}\frac{dX^\beta}{d\tau}\frac{dX^\gamma}{d\tau}, \tag{4.4}
$$

needs to be re-formulated in terms of $T \equiv TCG$. By performing successive applications of the chain rule we obtain a modified geodesic equation given by (Poisson and Will 2014)

$$
\begin{aligned}
\frac{d^2X^\alpha}{dT^2} &= \left(\frac{dT}{d\tau}\right)^{-1}\frac{d}{d\tau}\left[\left(\frac{dT}{d\tau}\right)^{-1}\frac{dX^\alpha}{d\tau}\right], \\
&= -\left(\Gamma^\alpha_{\beta\gamma} - \frac{V^\alpha}{c}\Gamma^0_{\beta\gamma}\right)V^\beta V^\gamma,
\end{aligned}
\tag{4.5}
$$

where we have defined the local geocentric coordinate velocity as $V^\alpha \equiv (c, V^j)$ where $V^j = dX^j/dT$ and we have made use of the fact that $X^0 = cT$. We note that (4.5) is zero for the case $\alpha = 0$, such that, the motion of a near-Earth satellite is determined purely by its spatial components given by (Soffel 2014)

[4]TCG is translated as Temps-Coordonnée Géocentrique and is the French acronym for geocentric coordinate time.

$$\frac{d^2X^i}{dT^2} = -c^2\left[\Gamma^i_{00} + 2\Gamma^i_{0j}\frac{V^j}{c} + \Gamma^i_{jk}\frac{V^j}{c}\frac{V^k}{c} - \left(\Gamma^0_{00} + 2\Gamma^0_{0j}\frac{V^j}{c} + \Gamma^0_{jk}\frac{V^j}{c}\frac{V^k}{c}\right)\frac{V^i}{c}\right]. \tag{4.6}$$

Finally, the equations of motion for a near-Earth satellite are obtained by substituting the appropriate Christoffel symbols in (4.6) and are given by

$$\frac{d^2X^j}{dT^2} = \partial_j W + \frac{1}{c^2}\left[\left(V^2 - 4\,W\right)\partial_j W - 4\,V^j V^k \partial_k W - 4(\partial_j W^k - \partial_k W^j)V^k\right]. \tag{4.7}$$

We emphasise again that we make no assumptions concerning the four gravitational potentials, other than that they depend on time implicitly through their spatial coordinates, and otherwise denote arbitrary spatially dependent solutions of Poisson equations.

4.5 Energy Integral

In classical Newtonian mechanics, there exist a number of first integrals associated with Newton's laws of motion, arising from the conservation laws of energy, linear momentum and angular momentum. Conservation of energy usually arises from Newton's second law simply by multiplication of the velocity components and integration to deduce a first integral sometimes referred to as Jacobi's integral (Goldstein 1965).

Following an identical formal procedure, there exists a corresponding energy integral in the post-Newtonian regime associated with (4.7). This is achieved by multiplying (4.7) by $V^j = dX^j/dT$ such that

$$\frac{d}{dT}\left(\frac{V^2}{2} - W + \frac{2W^2}{c^2}\right) = \frac{1}{c^2}\left(V^2\frac{dW}{dT} - 4V^2\frac{dW}{dT} - 4V^k\frac{dW^k}{dT} + 4\frac{dW^j}{dT}V^j\right), \tag{4.8}$$

where we have used $V^j\partial_j W \equiv dW/dT$ which is a consequence of assuming the gravitational field is stationary. Now since the final two terms cancel, Eq. (4.8) can be equivalently expressed by

$$\frac{d}{dT}\left(\frac{V^2}{2} - W + \frac{2W^2}{c^2}\right) = -\frac{3V^2}{c^2}\frac{dW}{dT}. \tag{4.9}$$

It can be readily established that on formally dividing (4.9) by dW/dT, we obtain a first order, linear ordinary differential equation for V^2 as a function of W, which may be solved to obtain

$$V^2 = \frac{5c^2}{9} - \frac{4W}{3} + Ce^{-6W/c^2}, \tag{4.10}$$

where C denotes an arbitrary constant of integration. From Eq. (4.10) it is apparent that for near-Earth satellites, the square of the magnitude of the velocity vector depends only on the single scalar gravitational potential W and not on the vector gravitational potentials W^j. This is due to the fact that the vector potential acts identically to the magnetic potential in electromagnetism which contributes a Lorentz-like force of the form $\mathbf{v} \times \mathbf{B}$ where $\mathbf{B} = \nabla \times \mathbf{W}$ which is formally identical to the final term in (4.7). As in electromagnetism, it is a well known fact that this term does not contribute to the energy. Alternatively, we may deduce this by noticing the Lorentz-like force and velocity vector are perpendicular by definition. In general relativity, this feature is studied under the rubric gravitomagnetism.

We observe that since W is prescribed modulo an arbitrary constant, we may re-write Eq. (4.10) as

$$V^2 = C_1 - \frac{4W}{3} + C_2 e^{-6W/c^2}, \tag{4.11}$$

as an integral of Eq. (4.7) involving two arbitrary constants C_1 and C_2. We further observe that if there is the requirement that Eq. (4.11) tends to the Jacobi integral $V^2 = constant + 2W$ (O'Keefe 1957) in the limit W tending to zero, then the constant $C_2 = -5c^2/9$ and Eq. (4.11) becomes

$$V^2 = C_1 - \frac{4W}{3} - \frac{5c^2}{9} e^{-6W/c^2}. \tag{4.12}$$

Hence, the post-Newtonian approximation to the classical Jacobi-integral is given by

$$V^2 \simeq C_1 - \frac{5c^2}{9} + 2W - \frac{10\,W^2}{c^2}, \tag{4.13}$$

where we have used the fact that for a near-Earth object $W/c^2 \ll 1$ and performed an expansion of the exponential term in (4.12) up to the same order as (4.1).

4.6 Variational Principle

In the previous section, we have derived a new energy integral (4.10) using algebraic manipulation of the post-Newtonian equations of motion (4.7). Equation (4.10) corresponds to energy conservation in the post-Newtonian approximation for a one body system, and there exists a corresponding Lagrangian which can be obtained from a variational principle.

There exists a number of integrals of motion corresponding to symmetries in a given spacetime, and we identify conservation of energy by the homogeneity of time, which is directly related to Noether's theorem (Goldstein 1965; Franklin 2010; Landau and Lifshitz 2013). The energy integral is derived from the variational principle which takes the form

$$\delta S = \delta \int \left(\frac{\mathrm{d}s}{\mathrm{d}t} \right) \mathrm{d}t = 0, \tag{4.14}$$

where the action is given by S and the associated Lagrangian $\mathcal{L}(X^j, V^j)$ is given by (Soffel 1989; Franklin 2010; Straumann 2004)

$$\mathcal{L} \equiv \left(\frac{\mathrm{d}s}{\mathrm{d}t} \right) = \left[-g_{\mu\nu} V^\mu V^\nu \right]^{1/2}, \tag{4.15}$$

noting that $(\mathrm{d}s)^2$ is the invariant spacetime interval associated with the post-Newtonian metric components and the implied summation here is over four indices.

The Lagrangian can be expressed as the sum of the Newtonian and post-Newtonian contributions given by

$$\mathcal{L} = \mathcal{L}_N + \mathcal{L}_{PN}, \tag{4.16}$$

where by definition (see Richardson and Kelly 1988 and references therein)

$$\frac{\mathcal{L}_{PN}}{\mathcal{L}_N} \sim \frac{1}{c^2}. \tag{4.17}$$

Making the appropriate substitutions in Eq. (4.15) with the metric tensor components (4.1) yields

$$\mathcal{L} = \left[1 - \frac{2W}{c^2} + \frac{2W^2}{c^4} + \frac{8W^j}{c^4} V^j - \frac{V^j V^j}{c^2} - \frac{2W}{c^4} V^j V^j \right]^{1/2}, \tag{4.18}$$

which is equivalently expressed as

$$\begin{aligned} \mathcal{L} = c^{-2} \bigg(& c^2 - W + \frac{W^2}{c^2} + 4 \frac{W^j V^j}{c^2} - \frac{V^j V^j}{2} - \frac{W}{c^2} V^j V^j - \frac{W^2}{2c^2} \\ & - \frac{W}{2c^2} V^j V^j - \frac{1}{8c^2} (V^j V^j)^2 \bigg), \end{aligned} \tag{4.19}$$

where we have expanded the square root in (4.18) to post-Newtonian order. The final expression for the Lagrangian is obtained by absorbing the leading constant and scaling the Lagrangian by $-c^{-2}$ (Richardson and Kelly 1988) which is given by

$$\mathcal{L} = \frac{V^j V^j}{2} + W - \frac{W^2}{2c^2} - 4 \frac{W^j V^j}{c^2} + \frac{3W}{2c^2} V^j V^j + \frac{(V^j V^j)^2}{8c^2}. \tag{4.20}$$

The equations of motion associated with $\mathcal{L}$ are obtained from the Euler–Lagrange equations given by

$$\frac{d}{dT} \left(\frac{\partial \mathcal{L}}{\partial V^j} \right) - \frac{\partial \mathcal{L}}{\partial X^j} = 0, \tag{4.21}$$

which are given explicitly by

$$\frac{d}{dT}\left(V^j\delta_{ij} - \frac{4}{c^2}W^j\delta_{ij} + \frac{3W}{c^2}V^j\delta_{ij} + \frac{1}{2c^2}(V^kV^k)V^j\delta_{ij}\right) - \frac{\partial W}{\partial X^i} + \frac{W}{c^2}\frac{\partial W}{\partial X^i}$$
$$+\frac{4}{c^2}\frac{\partial W^i}{\partial X^j}V^j - \frac{3}{2c^2}\frac{\partial W}{\partial X^i}(V^kV^k) = 0. \tag{4.22}$$

After some simplification, Eq. (4.22) reduces to

$$\frac{dV^i}{dT} + \frac{4W}{c^2}\frac{\partial W}{\partial X^i} - \frac{\partial W}{\partial X^i} + \frac{4}{c^2}\left(\frac{\partial W^j}{\partial X^i} - \frac{\partial W^i}{\partial X^j}\right)V^j = -\frac{3}{c^2}\frac{\partial W}{\partial X^i}(V^kV^k), \tag{4.23}$$

where we have substituted $dV^i/dT = \partial W/\partial X^i$ in to the post-Newtonian terms in Eq. (4.23) and upon multiplication by V^i, Eq. (4.23) simplifies and can be expressed as (4.8) so that Eq. (4.10) follows immediately.

We note that by following the same formal procedure of multiplication of the Euler–Lagrange equations of motion by the velocity vector, it is possible to derive the Legendre transformation relating the Hamiltonian and Lagrangian (Goldstein 1965). Formally, this procedure is equivalent to calculating the total time derivative of the Lagrangian $d\mathcal{L}/dT$ and is given by

$$V^j\frac{d}{dT}\left(\frac{\partial\mathcal{L}}{\partial V^j}\right) - V^j\frac{\partial\mathcal{L}}{\partial X^j} = 0, \tag{4.24}$$

where we note the term on the right is zero because the Lagrangian is time independent. Finally, by performing integration by parts on the first term on the left and remembering that $V^j = dX^j/dT$ yields

$$V^j\frac{\partial\mathcal{L}}{\partial V^j} - \mathcal{L} = \mathcal{E}, \tag{4.25}$$

where we identify $\mathcal{E}$ as the energy function (Goldstein 1965). When expressed explicitly, we may deduce that the energy given by (4.25) can be shown to be equivalent to the linear approximation of (4.10). It should be noted that this approach would require a series representation of the exponential function and can thought of as the approximate energy integral. The conditions for which they are equivalent are given below.

Without any assumptions as to the nature of the scalar W or the vector potentials $W^j (j = 1, 2, 3)$ we have derived, using algebraic manipulation, the Jacobi-like integral in the post-Newtonian regime. The derivations of the first post-Newtonian energy integral found in the literature (see for example Soffel 1989; Poisson and Will 2014) assume an explicit form of the potential. However, this is not the case above which is valid for any potential W satisfying the Poisson-like equations (4.2). Further, the energy integral (Eq. (4.13) or (4.27) depending on how the constant of integration is defined), under the appropriate transformation between local and global frames given in Damour et al. (1991), is equivalent to the conserved energy of the n-body problem associated with the Einstein–Infeld–Hoffmann equations (see Eq. $(3.2)_a$ in Will (2014)). Finally, we comment that (4.13) can be seen to coincide with Eq. (10.3) in Poisson and Will (2014) for the energy of a two-body system. Poisson and Will (2014) derive the energy integral from a complicated ad hoc dimensional derivation, by seeking an energy integral which is a linear combination of possibly six dimensionally consistent quantities. Following this, the derived expression is determined by a trial and error process (for further details we refer the reader to Poisson and Will 2014 page 482).

The derivation presented above is far more direct and straightforward, and most importantly incorporates the full exponential term which cannot be determined by the approach adopted in Poisson and Will (2014). To show (4.13) coincides with (10.3) of Poisson and Will (2014) we must take the limit $\eta \to 0$ so that (10.3) corresponds to the post-Newtonian energy of a test particle in a one-body system, where η is defined in Poisson and Will (2014) (page 482). On making the substitution $v^2 = 2(\mathcal{E} + U)$ in the post-Newtonian terms of (10.3) in Poisson and Will (2014), we obtain to post-Newtonian order

$$v^2 = 2U + 2\mathcal{E}(1 - 6U/c^2) - 10U^2/c^2. \tag{4.26}$$

Using the definition of energy at infinity and taking appropriate limits $W \to 0$ and $v^2 \to 2\mathcal{E}$ in (4.10) we can identify C in an alternative form given by $C = 2\mathcal{E} - 5c^2/9$ and hence express (4.10) as

$$V^2 = \frac{5c^2}{9}\left(1 - e^{-6W/c^2}\right) - \frac{4}{3}W + 2\mathcal{E}e^{-6W/c^2}. \tag{4.27}$$

Remembering $\mathcal{E} \sim W$ and expanding the above exponential functions to $O(W^2)$ and $O(W)$ respectively, the resulting expression coincides with (4.26).

Finally, we note that by following the same formal procedure outlined above it can be shown that Eq. (4.26) follows immediately from explicit calculation of (4.25).

4.7 Numerical Simulation of Post-Newtonian Equations of Motion

In this section, we perform a numerical simulation in order to confirm the energy integral (4.12). In order to provide an independent confirmation of Eq. (4.12) and the determination of C_2, we employ a standard 4th order Runge–Kutta scheme and numerically integrate the full post-Newtonian equations of motion (4.7) subject to the potentials (4.28) using the mathematical software Mathematica. Up to this point, the only assumptions made regarding the scalar and vector potentials are that they are time-independent and satisfy Poisson-like equations (4.2), and are otherwise

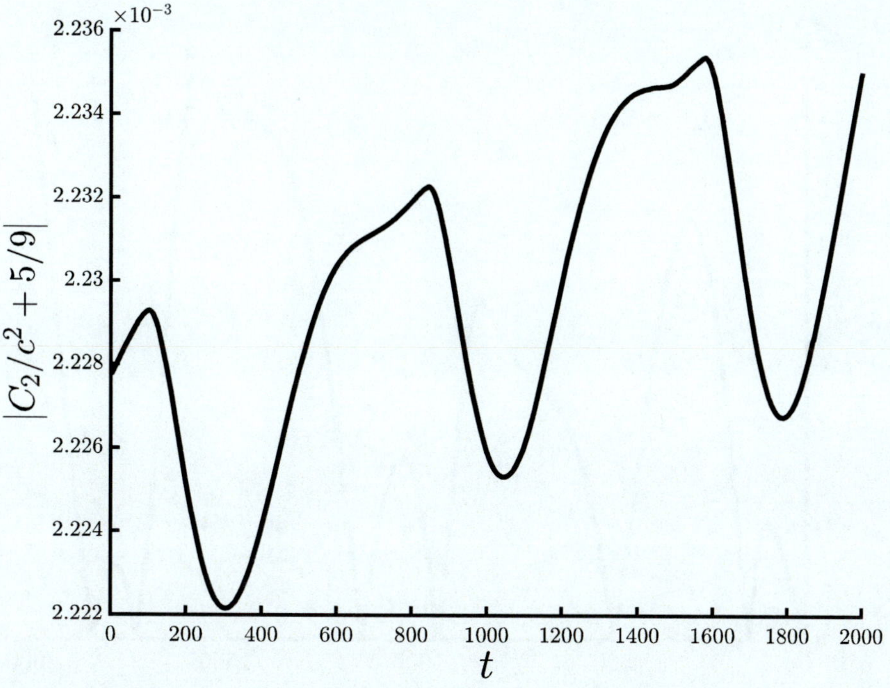

Fig. 4.1 Confirmation that $|C_2/c^2 + 5/9| \simeq 0$ (O'Leary et al. 2018)

arbitrary. For the numerical integration, we prescribe scalar and vector potentials according to the following expressions (Soffel and Langhans 2012)

$$W = GM/r, \qquad W_j = G\epsilon_{jik} S^i X^k / 2r^3, \tag{4.28}$$

where the magnitude of the position vector, the angular momentum of the Earth, the Earth's gravitational parameter, Newton's gravitational constant and the fully anti-symmetric Levi-Civita symbol are given by $r = |X^j|$, S^i,GM, G and ϵ_{jik} respectively.

Physically, the system described by (4.28) corresponds to a spherically symmetric, rotating model Earth (Soffel 1989). Standard numerical values are employed for $|S^i|$ and GM and are given by (Petit and Luzum 2010; Vallado 2001) $|S^i| = 9.8 \times 10^2$ km^2/s, and $GM = 0.398600 \times 10^6$ km^3/s^2, and the initial position (km) and velocity (km/s) used in the numerical simulation are obtained from NASA's Jet Propulsion Laboratory[5] for the LAGEOS spacecraft. The results of these simulations confirm that the value obtained for C_2/c^2 is within 10^{-3} of $-5/9$ as shown in Fig. 4.1 and the magnitude of C_2 remains constant up to a few parts-per-million (i.e. at the

[5]https://ssd.jpl.nasa.gov/.

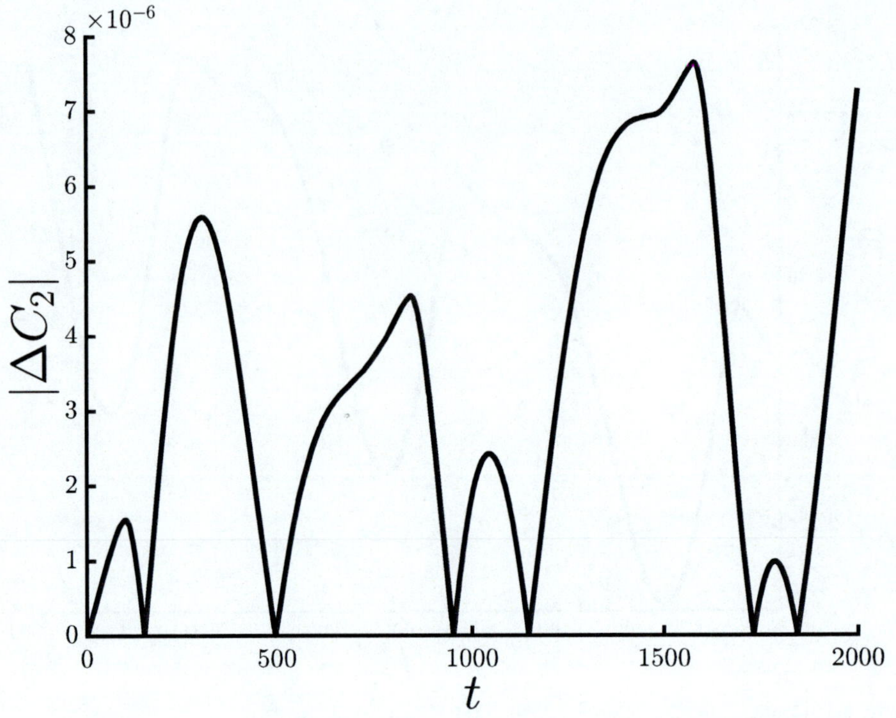

Fig. 4.2 Constancy of C_2. Variations in the absolute value of $|\Delta C_2|$ are only evident at the 10^{-6} level and otherwise remains constant

10^{-6} level) which is demonstrated in Fig. 4.2. Finally, we note the values for C_2 were determined from Eq. (4.11) given by

$$C_2 = \left(V_i^2 - V_j^2 + \frac{4}{3}(W_i - W_j)\right)\left(e^{-6W_i/c^2} - e^{-6W_j/c^2}\right)^{-1}, \quad (i, j = 1, 2, \dots, m), \tag{4.29}$$

where the subscripts i, j denote the ith and jth values of the position and velocity respectively which are obtained using the numerically simulated data.

4.8 Conservation of Post-Newtonian Energy Integrals

In this section, we compare the long-term evolution of energy variation for the newly derived energy integrals (4.12) and (4.27) with the corresponding linear approximations (4.31) and (4.33) which appear in the literature (Poisson and Will 2014). For potential $W = GM/r$ these are given explicitly by

$$V^2 = C_1 - \frac{4}{3}\left(\frac{GM}{r}\right) - \frac{5c^2}{9}\exp\left(-\frac{6GM}{c^2r}\right), \tag{4.30}$$

$$V^2 = C_1 - \frac{5c^2}{9} + 2\left(\frac{GM}{R}\right) - 10\left(\frac{GM}{cR}\right)^2, \tag{4.31}$$

and

$$V^2 = \frac{5c^2}{9}\left(1 - \exp\left(-\frac{6GM}{c^2r}\right)\right) - \frac{4}{3}\left(\frac{GM}{r}\right) + 2\mathcal{E}\exp\left(-\frac{6GM}{c^2r}\right), \tag{4.32}$$

$$V^2 = 2\left(\frac{GM}{r}\right) - 10\left(\frac{GM}{cr}\right)^2 + 2\mathcal{E}\left(1 - 6\frac{GM}{c^2r}\right), \tag{4.33}$$

respectively.

In general, classical numerical methods for solving differential equations such as the well-known 4th order Runge–Kutta scheme do not preserve first integrals such as energy and angular momentum. This leads to a dissipative-like effect giving incorrect qualitative behaviour over long integration periods. Symplectic integration schemes (Hairer et al. 2006; McLachlan and Quispel 2006) have been successfully developed to combat such dissipative effects associated with classical schemes providing almost near conservation of energy integrals. Applications of symplectic integration schemes arise in areas such as molecular dynamics, particle accelerator physics and celestial mechanics (see Hairer et al. 2006; McLachlan and Quispel 2006 and references therein).

Symplectic integrators are aptly named due to their ability to preserve the symplectic structure associated with Hamiltonian systems (Donnelly and Rogers 2005; McLachlan and Quispel 2006) and provide an opportunity to compare the long-term evolution of the invariant quantities (4.30) and (4.32) with their associated approximations (4.31) and (4.33) respectively. Figure 4.3 demonstrates the advantage of using symplectic integration schemes by comparing the long-term drift in energy using a second-order Gauss–Legendre scheme (symplectic) with a 4th-order explicit Runge–Kutta. It is shown that the long-term error in energy ΔE associated with (4.30) remains bounded and close to the true value when using symplectic schemes; and shows non-linear secular drifts when using a 4th-order Runge–Kutta.

The post-Newtonian literature assumes the energy integrals are sufficiently represented by the corresponding Taylor series approximations. This gives rise to the important question as to whether there is any benefit in retaining the exponential functions in describing the dynamics associated with bodies in the post-Newtonian regime. The answer is in the affirmative and is demonstrated in the following section.

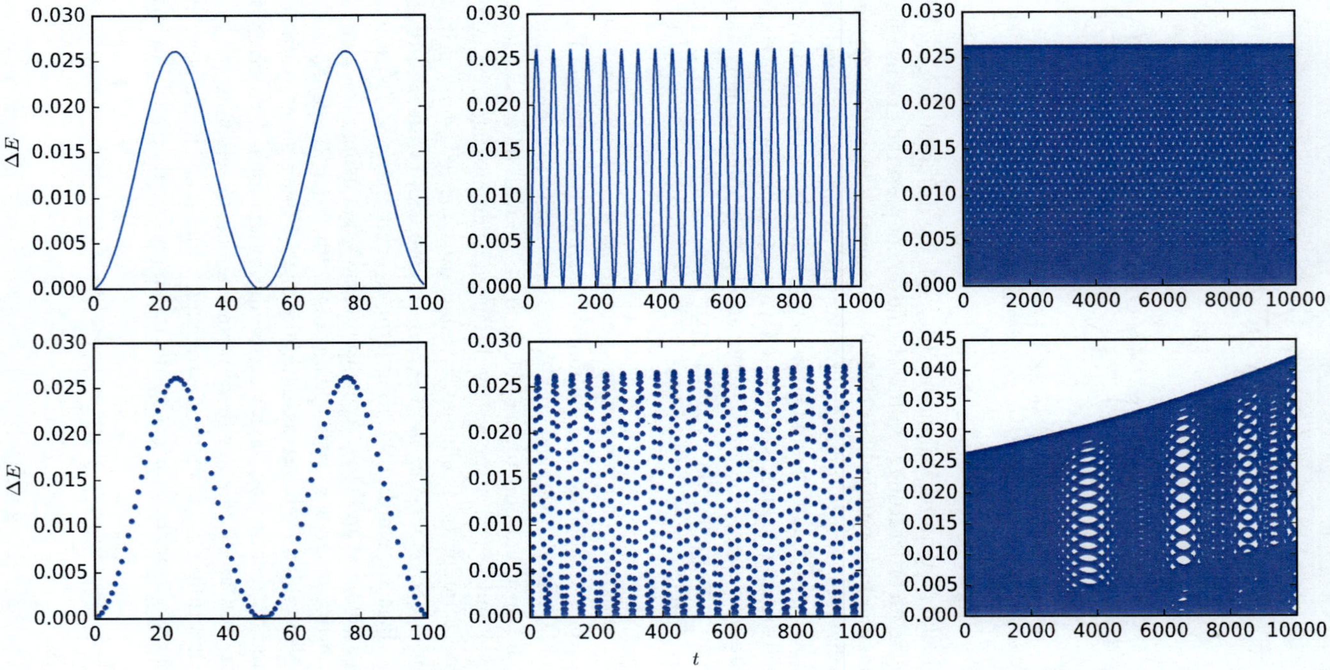

Fig. 4.3 Long-term conservation of exponential energy integral (4.30). **Top row:** Bounded energy associated with second-order Gauss Legendre symplectic integrator. **Bottom row:** Non-linear secular drift in energy associated with 4th order Runge–Kutta scheme

4.9 Numerical Methods

Introducing dimensionless variables according to $x^j = X^j/X_c$ and $t = T/T_c$ where X_c, T_c are characteristic units given by $X_c = GM/c^2$ and $T_c = GM/c^3$ respectively greatly simplifies the first post-Newtonian equations of motion computationally. Hence, the non-dimensionalised equations of motion of a test particle for the static and stationary Schwarzschild field of the Earth are given by

$$\frac{d^2x^i}{dt^2} = -\frac{x^i}{r^3} + 4\frac{x^i}{r^4} - v^2\frac{x^i}{r^3} + 4\frac{v^i}{r^3}\left(x_j v^j\right). \tag{4.34}$$

To demonstrate the difference between retaining the exponential functions in (4.30) and (4.32) over the corresponding Taylor series approximations (4.31) and (4.33), we perform several simulations by numerically integrating the non-dimensionalised post-Newtonian equations of motion (4.34) with a second-order Gauss–Legendre integration scheme. We note that all simulations are subject to the same initial conditions (appropriately scaled for the simulation) as Sect. 4.7 and are given explicitly by $x^i(0) = (3.6, 2.3, 1.1)$, $v^i(0) = (-0.004, -0.003, 0.0002)$, with integration time step $h = 0.01$. Figures 4.4 and 4.5 demonstrate the error in magnitude is greater

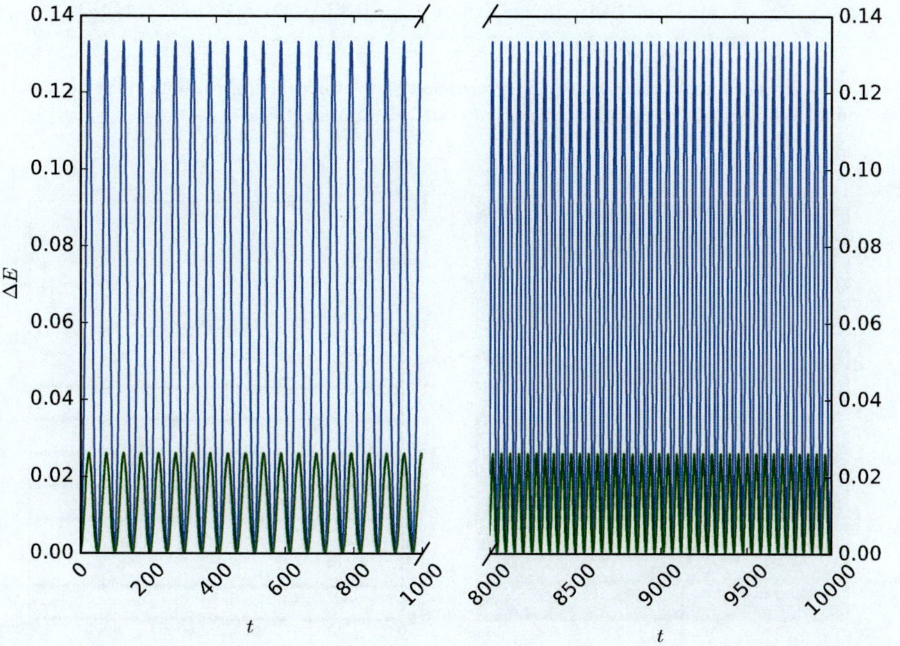

Fig. 4.4 Error in energy integrals (4.30) (green) and (4.31) (blue). Errors associated with (4.30) and (4.31) are bounded due to nature of symplectic schemes. Magnitude of error is larger for Taylor series approximation

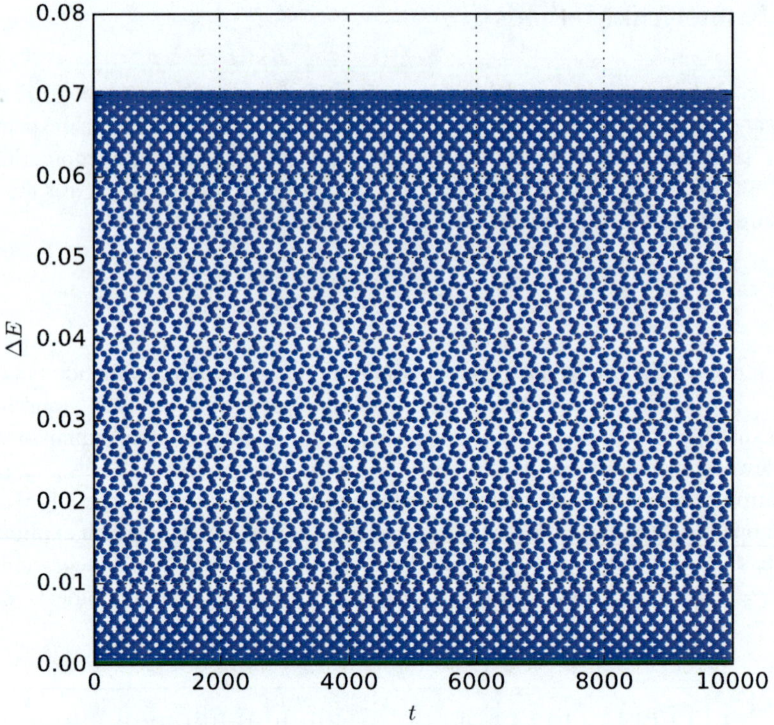

Fig. 4.5 Error in energy integrals (4.32) (green) and (4.33) (blue). Error for both energy integrals is bounded. Magnitude of error associated with (4.32) is of order 10^{-10}

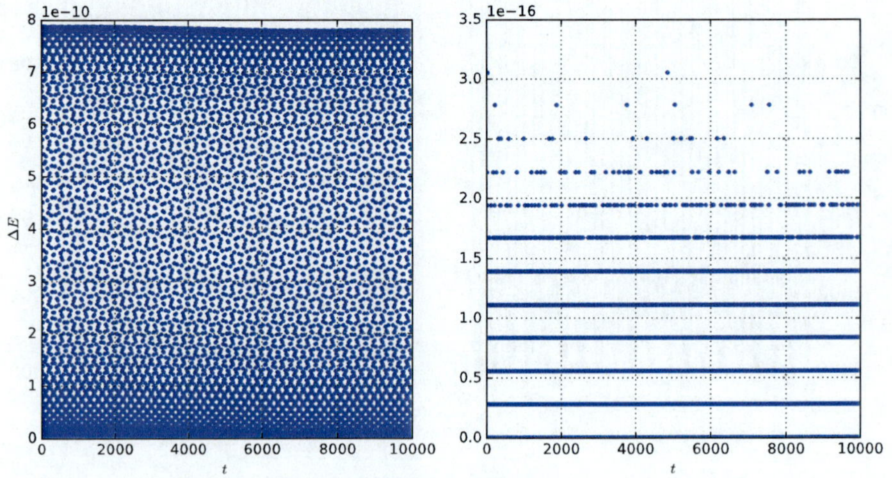

Fig. 4.6 Long-term conservation of exponential representation of energy integral (4.32). **Left:** Bounded error associated with Eq. (4.32). **Right:** Complete conservation (up to round-off error) of Eq. (4.32) using a symplectic 10th order implicit Runge–Kutta scheme

for both approximation methods (4.31) and (4.33) indicating that exponential representation of energy integrals given by (4.30) and (4.32) more closely resemble the conserved dynamics associated with a test particle in the post-Newtonian regime over long-term integration. Further, Fig. 4.6 indicates that the error in (4.32) is almost completely conserved with an error of the order 10^{-10}.

4.10 Conclusions

The post-Newtonian literature assumes the energy associated with a test particle is sufficiently represented by approximate forms given by Eqs. (4.31) and (4.33). By seeking an exact Jacobi-like integral in Sects. 4.5 and 4.6; an exponential term arises in the energy integral associated with a test particle. In this chapter, we have shown that the exponential term in the energy integrals (4.30) and (4.32) more closely resemble the true conserved dynamics of test particles in the post-Newtonian regime. The long-term magnitude in error for the Taylor series approximations (4.31) and (4.33) is shown to be larger in both cases. Further, the long-term behaviour in error associated with Eq. (4.32) is shown to be of order 10^{-10}. It can be shown that the energy integral given by Eq. (4.32) is completely conserved (up to round-off error) by increasing the order of the symplectic scheme and is discussed in Fig. 4.6. The complete conservation of (4.32) requires further investigation.

Chapter 5
Post-Newtonian Satellite Orbits

5.1 Introduction

As early as 1916, Einstein hypothesised (Einstein 1916a, b; 1918) the existence of gravitational waves using a weak-field approximation (d'Inverno 1992; Misner et al. 1973; Weinberg 1972) of the field equations of gravity. The first direct observation of gravitational waves was due to the coalescence of a binary black hole system (Abbott et al. 2016), which was detected by the advanced laser interferometer gravitational-wave observatory collaboration (Abbott et al. 2009; Harry and Collaboration 2010). More recently, a simultaneous observation of both gravitational waves (Abbott et al. 2017) *and* electromagnetic radiation (Goldstein et al. 2017) due to the merger of a binary neutron star system was detected by the collaboration and the low Earth orbit *Fermi*—Gamma-ray burst monitor (Meegan et al. 2009) respectively.

The treatment of such binary systems under the full non-linear field equations of general relativity requires a given spacetime to provide a complete description of the entire history of the binary system. Specifically, the solution is required to describe past events such as the formation of the bodies, and future events such as the final merger of the binary system (Poisson and Will 2014). Due to the complexity of the field equations, no such solution presently exists. Hence, to accommodate the complexities associated with Einstein's non-linear theory of gravity, we must exploit frameworks such as the DSX post-Newtonian formalism and emerging fields such as numerical relativity (Baumgarte and Shapiro 2010).

The n-body relativistic equations of motion are given by the Einstein–Infeld–Hoffmann equations. For $n = 2$, we investigate the associated dynamics of two-body systems in the first post-Newtonian approximation. The present chapter derives a new Binet equation in the first post-Newtonian regime which characterises the orbit of test particles. The post-Newtonian Binet equation is obtained by direct integration of the associated planar equations of motion. The newly obtained expression contains exponential functions that when approximated can produce known linear results.

J. O'Leary, *General Relativistic and Post-Newtonian Dynamics for Near-Earth Objects and Solar System Bodies*, Springer Theses,
https://doi.org/10.1007/978-3-030-80185-4_5

The accuracy of the post-Newtonian Binet equation is verified by comparing the long-term secular variations of both orbit equations (5.16) and (5.22) with the full non-linear general relativistic orbit equation (5.25). The results of the numerical simulation suggest that retaining the exponential functions produces orbits which closely resemble that of general relativity.

The chapter is organised as follows: in Sect. 5.2, we outline the procedure required to derive the relativistic n-body equations of motion associated with non-spinning point particles. In Sect. 5.3, we derive a new post-Newtonian Binet equation associated with the relativistic Kepler problem offering insight in to the mechanics of orbiting bodies in the relativistic domain for both near-Earth objects and astronomical binary systems which are currently of significant interest to astronomical communities. The expression obtained confirms the same formal mathematical structure as the energy integral derived in Chap. 4 and in Sect. 5.4, we provide an independent verification of the post-Newtonian Binet equation using the associated first integrals. Finally, the post-Newtonian Binet is compared with the general relativistic analogue in Sect. 5.5 and some numerical results are presented.

The following chapter and the results discussed have been published in O'Leary and Hill (2018)

- O'Leary, J. and Hill, J.M., 2018. Post-Newtonian satellite orbits. *Astrophysics and Space Science*, 363(10), p. 211.

5.2 The Einstein–Infeld–Hoffmann Equations

The complete treatment of the translational and rotational equations of motion for the general relativistic n-body problem describing arbitrarily shaped, rotating and deformable bodies is given in Damour et al. (1991, 1992, 1993, 1994). The simplest example of the relativistic n-body equations of motion is provided by the monopole model where it is assumed bodies are sufficiently separated, non-spinning and produce spherically symmetric gravitational fields and all tidal effects are ignored.

In this section, we derive the Einstein–Infeld–Hoffmann equations for the truncated monopole model as detailed in Damour et al. (1991) and Soffel (2014). This is achieved by introducing global external scalar and vector potentials $\bar{w}^A$ and $\bar{w}_j^A$ and metric tensor components $\bar{g}_{\alpha\beta}^A$ attached to some body $A = 1, 2, \ldots, N$ through the relations (Damour et al. 1991)

$$\bar{w}^A = \sum_{B\neq A} w^B, \qquad \bar{w}_j^A = \sum_{B\neq A} w_j^B, \tag{5.1}$$

and

$$\bar{g}^A_{00} = -\exp\left(-\frac{2}{c^2}\sum_{B\neq A} w^B\right),$$
$$\bar{g}^A_{0j} = -\frac{4}{c^3}\sum_{B\neq A} w^B_j, \tag{5.2}$$
$$\bar{g}^A_{ij} = \delta_{ij}\exp\left(\frac{2}{c^2}\sum_{B\neq A} w^B\right),$$

respectively. If we denote the position of body A by $x^i_A \equiv \mathbf{x}_A$, then the translational equations of motion for body A due to the remaining $n-1$ bodies are given explicitly by (Damour et al. 1991)

$$\frac{\mathrm{d}^2 x^i_A}{\mathrm{d}t^2} = \left[1 - \frac{4}{c^2}\bar{w}_A + \frac{1}{c^2}\left(v^A_j v^j_A\right)\right]\partial_i \bar{w}_A + \frac{4}{c^2}\partial_t \bar{w}^A_i - \frac{4}{c^2}\left(\partial_i \bar{w}^A_j - \partial_j \bar{w}^A_i\right) v^j_A$$
$$- \frac{1}{c^2}\left(3\partial_t \bar{w}_A + 4 v^j_A \partial_j \bar{w}_A\right) v^i_A, \tag{5.3}$$

and we denote the global coordinate velocity of body A by $v^i_A = \mathrm{d}x^i_A/\mathrm{d}t$. Further, we remark the global translational equations of motion for body A are derived by substituting the Christoffel symbols (4.3) in the modified geodesic equation (4.5). The detailed and exhaustive treatment of relativistic reference frames provided by the DSX formalism allows us to easily determine an explicit expression for the external potentials.

In the local frame of body B the scalar and vector potentials attached to a spherically symmetric, non-rotating body are given by

$$W^B = \frac{GM_B}{R_B}, \qquad W^B_j = 0. \tag{5.4}$$

On applying the transformation law between global and local frames (see Eqs. (4.12a) and (4.12b) of theorem 3 in Damour et al. (1991)) and making use of the formulae relating the inverse distance of a given body to an arbitrary reference point, we find w^B and w^B_j are given by Soffel (2014)

$$w^B = \frac{GM_B}{r_B}\left\{1 + 2\frac{\mathbf{v}^2_B}{c^2} - \frac{\bar{w}^B}{c^2} - \frac{1}{2c^2}\left[(\mathbf{v}_B\cdot\mathbf{n}_B)^2 + (\mathbf{a}_B\cdot\mathbf{r}_B)\right]\right\}, \qquad w^B_j = \frac{GM_B}{r_B} v^j_B, \tag{5.5}$$

where $\bar{w}^B$ denotes the external potential for body B and is equivalent to Eq. (5.1) with an appropriate change of label and $\mathbf{r}_B \equiv \mathbf{x} - \mathbf{x}_B$, $\mathbf{n}_B \equiv \mathbf{r}_B/|\mathbf{r}_B|$. The Einstein–Infeld–Hoffmann equations are obtained by substituting Eqs. (5.5) in (5.3) and are given by (Soffel 2014; Damour et al. 1991)

$$
\begin{aligned}
\frac{\mathrm{d}^2 x_A^i}{\mathrm{d}t^2} = &- \sum_{B\neq A} \frac{GM_B}{r_{AB}^2} \hat{\mathbf{r}}_{AB} \Bigg\{ 1 + c^{-2} \Bigg[\mathbf{v}_A^2 + 2\mathbf{v}_B^2 - 4\left(\mathbf{v}_A \cdot \mathbf{v}_B\right) - \frac{3}{2}\left(\hat{\mathbf{r}}_{AB} \cdot \mathbf{v}_B\right)^2 \\
&- 4 \sum_{C\neq A} \frac{GM_C}{r_{AC}} - \sum_{C\neq B} \frac{GM_C}{r_{BC}} \left(1 + \frac{1}{2} \frac{r_{AB}}{r_{CB}} \left(\hat{\mathbf{r}}_{AB} \cdot \hat{\mathbf{r}}_{CB} \right) \right) \Bigg] \Bigg\} \\
&+ \sum_{B\neq A} \left(\mathbf{v}_A - \mathbf{v}_B\right) \frac{GM_B}{c^2 r_{AB}^2} \left[4\left(\hat{\mathbf{r}}_{AB} \cdot \mathbf{v}_A\right) - 3\left(\hat{\mathbf{r}}_{AB} \cdot \mathbf{v}_B\right) \right] \\
&- \frac{7}{2} \sum_{B\neq A} \sum_{C\neq B} \frac{G^2 M_B M_C}{c^2 r_{AB} r_{BC}^2} \hat{\mathbf{r}}_{BC},
\end{aligned}
\tag{5.6}
$$

where $\mathbf{r}_{AB} \equiv \mathbf{r}_A - \mathbf{r}_B$ and $\hat{\mathbf{r}}_{AB} \equiv \mathbf{r}_A - \mathbf{r}_B / |\mathbf{r}_A - \mathbf{r}_B|$.

Finally, we note the Einstein–Infeld–Hoffmann equations of motion can be equivalently obtained from the n-body post-Newtonian Lagrangian which has been derived by many authors e.g. see Eq. (7.22) of Damour et al. (1991).

5.3 The Post-Newtonian Kepler Problem

The *Kepler* problem is the most well studied problem in classical celestial mechanics. It describes the motion of two spherically symmetric bodies about a common barycentre due to their mutual gravitational attraction. The general relativistic and Newtonian two-body problems are covered extensively in the literature (Blanchet et al. 2011; Damour 1987; Damour and Deruelle 1985), and the perturbed Kepler problem plays a central role in fields such as satellite orbit determination (Montenbruck and Gill 2012; Vallado 2001). In this section, we present a new formal solution to the first post-Newtonian Kepler problem and derive the associated post-Newtonian Binet equation (d'Inverno 1992) characterising the orbit of test particles.

The post-Newtonian two-body equations of motion are obtained from Eqs. (5.6) by calculating the relative acceleration $\mathbf{a} = \mathbf{a}_1 - \mathbf{a}_2$ and are given by (Poisson 2007; Poisson and Will 2014)

$$
\frac{d^2\mathbf{r}}{dt^2} = -\frac{\mu}{r^2}\hat{\mathbf{r}} - \frac{\mu}{c^2 r^2}\left[(1+3\eta)(\mathbf{v}\cdot\mathbf{v}) - \frac{3}{2}\eta\dot{r}^2 - 2(2+\eta)\frac{\mu}{r}\right]\hat{\mathbf{r}} + \frac{2\mu}{c^2 r^2}(2-\eta)\dot{r}\mathbf{v},
\tag{5.7}
$$

where we introduce $\mu = G(m_1 + m_2)$, $\eta = m_1 m_2/(m_1 + m_2)^2$ and we denote the mass of body i (for $i = 1, 2$) and the radial component of the velocity as m_i and $\dot{r}$ respectively. Finally, derivatives with respect to time are denoted using the usual dot notation. We note the dimensionless mass parameter η is bounded according to $0 \leq \eta \leq 1/4$ and in the limit $\eta \to 0$, we recover the one-body problem (Damour and Deruelle 1985).

Equation (5.7) is decomposed into a coupled system of scalar equations by introducing planar polar coordinates (r, θ) according to

$$\mathbf{r} = r\hat{\mathbf{r}}, \qquad \mathbf{v} = \dot{r}\hat{\mathbf{r}} + r\dot{\theta}\hat{\boldsymbol{\theta}}, \qquad \mathbf{a} = (\ddot{r} - r\dot{\theta}^2)\hat{\mathbf{r}} + (r\ddot{\theta} + 2\dot{r}\dot{\theta})\hat{\boldsymbol{\theta}}. \tag{5.8}$$

The corresponding system of coupled differential equations associated with (5.7) are given by (Poisson 2007; Poisson and Will 2014)

$$\begin{gathered}\ddot{r} = r\dot{\theta}^2 - \frac{\mu}{r^2} + \frac{\mu}{c^2 r^2}\left[3\dot{r}^2 - \frac{7}{2}\eta\dot{r}^2 - (1+3\eta)(r\dot{\theta})^2 + 2(2+\eta)\frac{\mu}{r}\right],\\ 2r\dot{r}\dot{\theta} + r^2\ddot{\theta} = 2(2-\eta)\frac{\mu}{c^2}\dot{r}\dot{\theta},\end{gathered} \tag{5.9}$$

and Eqs. (5.9) are separated in to respective radial and angular components. We note Eq. $(5.9)_b$ can be equivalently expressed in the convenient form given by

$$\frac{\ddot{\theta}}{\dot{\theta}} = 2(2-\eta)\frac{\mu}{c^2 r^2}\dot{r} - 2\frac{\dot{r}}{r}. \tag{5.10}$$

The integration of $(5.9)_b$ with respect to time produces an alternative expression for the angular momentum in the post-Newtonian Kepler problem in planar polar coordinates given by

$$r^2\dot{\theta} = h_0 e^{-2(2-\eta)\mu/(c^2 r)}, \tag{5.11}$$

where h_0 is a constant of integration.

The formal procedure adopted in celestial mechanics is to introduce a Binet transformation (d'Inverno 1992) where θ replaces time as the independent variable through the relations

$$r = \frac{1}{u}, \quad \dot{r} = -h_0 u' e^{-2(2-\eta)\mu u/c^2}, \quad \ddot{r} = -h_0^2 u^2\left[u'' - \frac{2(2-\eta)\mu}{c^2}(u')^2\right]e^{-4(2-\eta)\mu u/c^2}, \tag{5.12}$$

where $u = u(\theta)$ and $u' = du/d\theta$. The simplification of $(5.9)_a$ is greatly facilitated by multiplication of $-2u'/(uh_0)^2$ and substitution of the relations (5.12) given by

$$\begin{aligned}\left[2u'(u''+u) - \frac{\mu u'(2+3\eta)}{c^2}((u')^2+u^2)\right]e^{-4(2-\eta)\mu u/c^2} &= \frac{2\mu u'}{h_0^2}\\ - \frac{4\mu^2 u u'(2+\eta)}{h_0^2 c^2} + \frac{3\eta\mu u^2 u'}{c^2}e^{-4(2-\eta)\mu u/c^2}&.\end{aligned} \tag{5.13}$$

We note that in the test particle ($\eta \to 0$) and Newtonian limits ($c \to \infty$), Eq. (5.13) reduces to the well known classical Binet equation (d'Inverno 1992) with associated solution describing conic sections. A new method to derive the equivalent first post-Newtonian Binet equation is presented below.

By introducing $\delta \equiv (u')^2 + u^2$, we express Eq. (5.13) as a first order differential equation given by

$$\left(\delta e^{-(2+3\eta)\mu u/c^2}\right)' = \frac{2\mu u'}{h_0^2}\left[1 - \frac{2\mu u(2+\eta)}{c^2}\right]e^{(6-7\eta)\mu u/c^2} + \frac{3\eta\mu u^2 u'}{c^2}e^{-(2+3\eta)\mu u/c^2}, \tag{5.14}$$

where we note Eq. (5.14) has been multiplied by $\exp[(6 - 7\eta)\mu u/c^2]$ in order to obtain the appropriate integrating factor.

The formal analytical solution to Eq. (5.14) can be reduced to quadrature and would need to be evaluated numerically as an integral. The corresponding second order differential equations characterising orbits for binary and one-body systems in the first post-Newtonian regime are given by

$$u'' + \left(1 + \frac{3\eta}{2+3\eta} + \frac{8\mu^2(4-\eta^2)}{h_0^2c^2(6-7\eta)}e^{4(2-\eta)\mu u/c^2}\right)u =$$
$$\frac{\delta_0\mu(2+3\eta)}{2c^2}e^{(2+3\eta)\mu u/c^2} + \frac{2\mu(17\eta^2 - 32\eta + 28)}{h_0^2(6-7\eta)^2}e^{4(2-\eta)\mu u/c^2} - \frac{3\eta c^2}{\mu(2+3\eta)^2}, \tag{5.15}$$

and

$$u'' + \left(1 + \frac{16\mu^2}{3h_0^2c^2}e^{8\mu u/c^2}\right)u = \frac{14\mu}{9h_0^2}e^{8\mu u/c^2} - \frac{5\mu}{9h_0^2}e^{2\mu u/c^2}, \tag{5.16}$$

respectively. We note that δ_0 is introduced above as an arbitrary constant of integration arising from (5.14). Equation (5.16) is the complete analytical first post-Newtonian Binet equation. The value of δ_0 is determined by adding the requirement that in the Newtonian limit, we recover the classical Binet equation and is given by $\delta_0 = -5c^2/9h_0^2$. Again, we note, the value of δ_0 is consistent with results discussed in Chap. 4.

5.4 Derivation of Post-Newtonian Binet Equation from First Integrals

In both general relativity and classical Newtonian gravity, the corresponding orbit equations are derivable from direct algebraic manipulation of the equations of motion or by exploiting the conservation of angular momentum in the energy integral (both derivations are given in (B.1) and (B.2)). As an independent verification of the results presented here and in Chap. 4, we derive the post-Newtonian Binet equation by exploiting the orbital angular momentum in the energy integral (4.10).

The orbital angular momentum for a test particle in the first post-Newtonian regime is obtained following calculation of the vector cross product of (4.7) with the test particle position in the local frame X^j according to

$$\epsilon_{ijk}X^j\frac{d^2X^k}{dt^2} = \epsilon_{ijk}X^j\frac{\partial W}{\partial X^k} + \frac{1}{c^2}\left[(V^2 - 4W)\epsilon_{ijk}X^j\frac{\partial W}{\partial X^k} - 4\epsilon_{ijk}X^jV^k\left(V^l\frac{\partial W}{\partial X^l}\right)\right], \tag{5.17}$$

where the vector cross product is given by the Levi-Civita symbol ϵ_{ijk}. It is important to note that the gradient of the scalar potential W arising in (5.17) is equivalently expressed as $\partial W/\partial X^k e^k$ where e^k is a unit basis vector in the direction of the k-th coordinate. Hence, the vector cross product of the gradient of the scalar potential with the corresponding test particle position is zero due to the anti-symmetry associated with the exchange of any two indices of the Levi-Civita symbol. Equation (5.17) is then equivalently expressed in terms of the first order, vector differential equation according to

$$\frac{\mathrm{d}}{\mathrm{d}t}(\epsilon_{ijk}X^j V^k) + \frac{4}{c^2}\frac{\mathrm{d}W}{\mathrm{d}t}(\epsilon_{ijk}X^j V^k) = 0, \tag{5.18}$$

where the total time derivative of the scalar potential W is given by $\mathrm{d}W/\mathrm{d}t \equiv V^l \partial W/\partial X^l$. The corresponding solution to Eq. (5.18) follows immediately and is given by

$$\epsilon_{ijk}X^j V^k = h_i e^{-4W/c^2}, \tag{5.19}$$

where h_i is a constant vector following integration of (5.18). We note the approach presented above appears not to have been adopted in the literature.

Now, Eqs. (4.10) and (5.19) in planar polar coordinates are given by

$$\begin{aligned} \dot{r}^2 + r^2\dot{\theta}^2 &= \frac{5c^2}{9} - \frac{4W}{3} + Ce^{-6W/c^2}, \\ r^2\dot{\theta} &= h_0 e^{-4W/c^2}, \end{aligned} \tag{5.20}$$

where h_0 is denotes the magnitude of the constant vector h_i. By adopting the same formal procedure as Sect. 5.3 and prescribing the potential $W = \mu/r$ we find

$$\left[(u')^2 + u\right] h_0^2 e^{-8\mu u/c^2} = \frac{5c^2}{9} - \frac{4\mu}{3}u + Ce^{-6\mu u/c^2}, \tag{5.21}$$

and on differentiation with respect to the polar angle θ, it can be easily shown that Eq. (5.21) is equivalent to (5.16).

5.5 Discussion

The problem of motion in general relativity (see Damour 1987 for a thorough review) has a vast and extensive literature. Many authors have approached the two-body problem in the first-post-Newtonian approximation by formulating the problem in terms of osculating elements (see Kopeikin et al. 2011; Soffel 1989; Will 2011 and references therein). In contrast to this, we have developed the corresponding solutions by adopting a direct integration approach. In doing so, we obtain the fully analytical post-Newtonian Binet equation and as expected, in the appropriate limit we recover known results.

The post-Newtonian orbit equation (5.15) describes the motion of a binary system of arbitrary masses m_1 and m_2. Expanding the exponential functions (to post-Newtonian order) in (5.16) using a Taylor series approximation yields

$$u'' + \left(1 - \frac{6\mu^2}{h_0^2 c^2}\right) u = \frac{\mu}{h_0^2}. \tag{5.22}$$

The solution to (5.22) is well known and given by

$$u(\theta) = \frac{\mu}{h_0^2} + \frac{6\mu^3}{h_0^4 c^2} + u_0 \cos\left(\left[1 - \frac{6\mu^2}{h_0^2 c^2}\right]^{1/2} \theta - \theta_0\right), \tag{5.23}$$

where u_0, θ_0 are arbitrary constants of integration. The orbit described by Eq. (5.23) is no longer 2π-periodic but precesses about an angle

$$\Delta\theta = 2\pi \left(1 - \frac{6\mu^2}{h_0^2 c^2}\right)^{-1/2} \approx 2\pi + \frac{6\pi\mu^2}{h_0^2 c^2}. \tag{5.24}$$

Equation (5.24) is well-known and accounts for the relativistic contribution to planetary perihelion precession such as the anomalous precession associated with Mercury.

Equation (5.16) is the exact analytical first-post-Newtonian Binet equation and the same mathematical structure arises when deriving the energy associated with a test particle in the first-post-Newtonian approximation discussed in Chap. 4. With this, the question arises as to whether there is a benefit in retaining the exponential terms in (5.16) over the corresponding linear approximation (5.22). Again, the answer is in the affirmative and is presented below.

The general relativistic Binet equation is well known and presented in most general relativity undergraduate textbooks (d'Inverno 1992; Weinberg 1972) given by

$$u'' + u = \frac{\mu}{h_0^2} + \frac{3\mu}{c^2} u^2. \tag{5.25}$$

Equation (5.25) can be derived from a variational formulation and describes the geodesic motion of a test particle in Schwarzschild geometry (see Appendix B for derivation). Introducing $\xi = u/u_c$ in Eqs. (5.16), (5.22) and (5.25) we obtain the corresponding non-dimensionalised equations given by

$$\xi'' + \left(1 + \frac{16\beta}{3} e^{8\beta\xi}\right) \xi = \frac{14}{9} e^{8\beta\xi} - \frac{5}{9} e^{2\beta\xi}, \tag{5.26}$$

$$\xi'' + (1 - 6\beta)\,\xi = 1, \tag{5.27}$$

$$\xi'' + \xi = 1 + 3\beta\xi^2, \tag{5.28}$$

respectively where u_c is a characteristic unit with the same dimensions as u and $\beta \equiv \mu^2/(h_0^2 c^2)$ is a dimensionless parameter. To demonstrate the advantage of retaining the exponential functions in (5.16) we adopt the approach given in Koberlein and Meisel (2013) by identifying the long-term secular changes in the positions of (5.16)

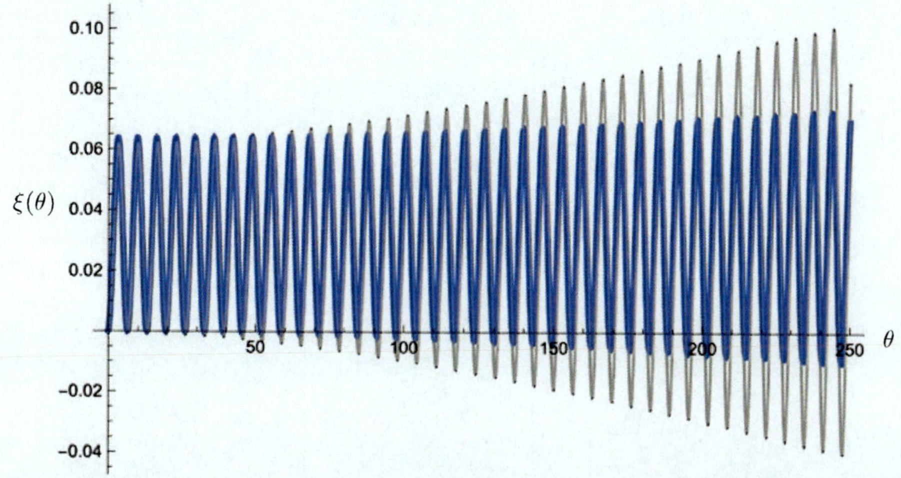

Fig. 5.1 Comparisons of secular variation in post-Newtonian orbits. At each integration step, we compare the full post-Newtonian orbit equation (5.26) (blue) and the linear approximation (5.27) (gray) with the full general relativistic Binet equation (5.28) by plotting the difference in positions ξ. The simulation uses initial conditions given by $\xi(0) = 1.3$, $\xi'(0) = 0$ and $\beta = 0.01$ which approximately correspond to an orbit with semi-major axis $a = 1$ and eccentricity $e = 0.5$ (O'Leary and Hill 2018)

and (5.22) when compared with the full general relativistic Binet equation (5.25) discussed in Fig. 5.1. We perform an orbit simulation by numerically integrating the non-dimensionalised differential equations given above, subject to the initial conditions $\xi(0) = 1.3$ and $\xi'(0) = 0$ and $\beta = 0.01$ which approximately correspond to an orbit with semi-major axis $a = 1$ and eccentricity $e = 0.5$. The results given in Fig. 5.1 suggest that the characterisation of relativistic orbits in the post-Newtonian regime should retain the exponential terms associated with the newly derived post-Newtonian Binet equation in order to closely resemble the orbit of a test particle in general relativity.

As a concluding remark to this chapter, we draw the reader' s attention to Figs. 5.2 and 5.3. The qualitative behaviour of the phase space trajectories varies largely as the value of β increases which is presented in Fig. 5.2. We note the solution to Eq. (5.27) is dependent on the coefficient of the linear term and for $\beta > 1/6$, the corresponding solution will be hyperbolic.

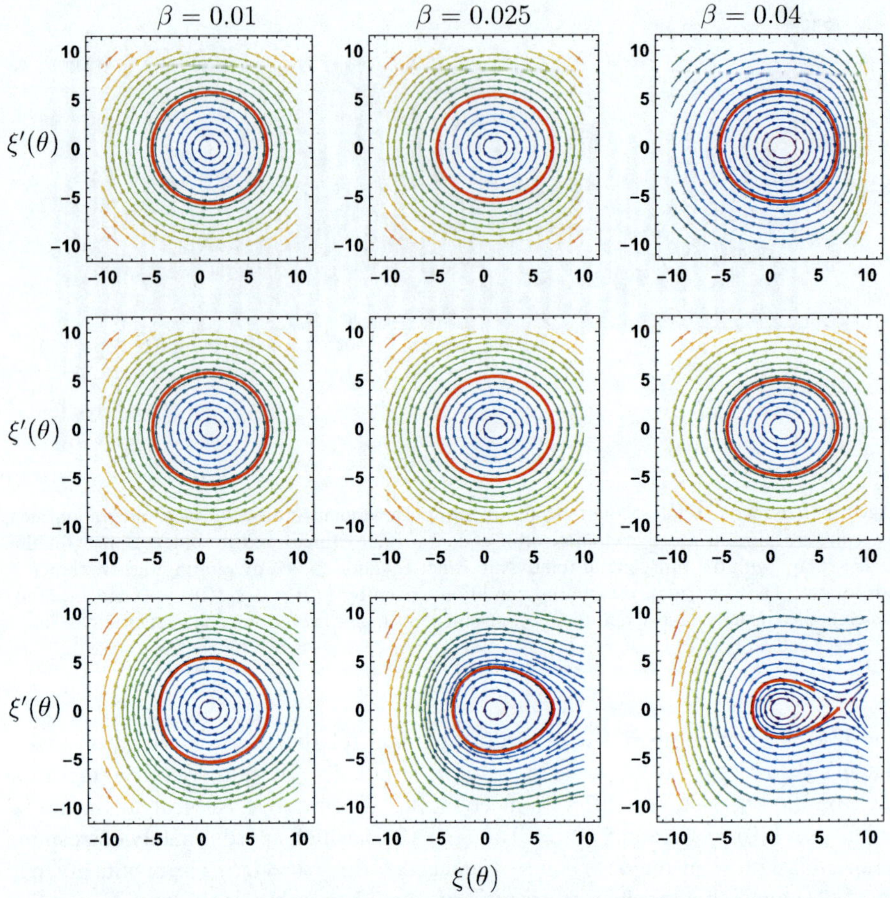

Fig. 5.2 Phase portrait diagrams of relativistic Binet equations. Typical phase space trajectories (plotted in red) for Eqs. (5.26) (top row), (5.27) (middle row) and (5.28) (bottom row) respectively. Trajectories are given for $\beta = 0.01, 0.025, 0.04$ respectively

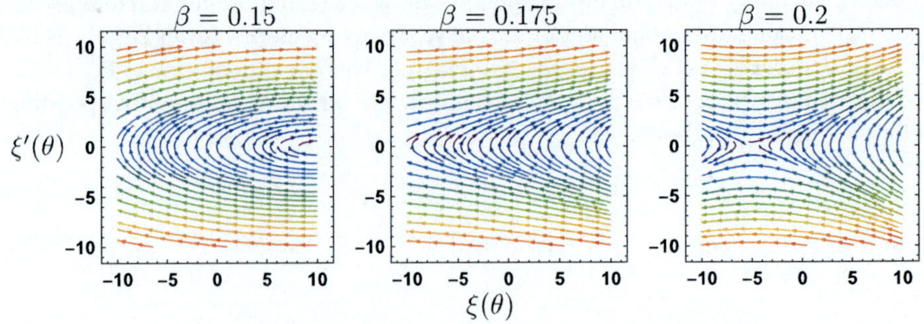

Fig. 5.3 Typical phase space trajectory associated with Eq. (5.27). Trajectories are given for $\beta = 0.15, 0.175, 0.2$ respectively

Chapter 6
Post-Newtonian Two-Centre Problem

6.1 Introduction

Fully integrable gravitational systems are not common in the literature and the solutions to many dynamical systems require advanced numerical methods. A class of dynamical systems which are integrable are those which are separable (Whittaker 1988). A further class of integrable systems discovered by Liouville (see Ó Mathúna 2008 for a detailed historical account) are the particular systems with kinetic and potential energies prescribed according to Eqs. (6.11). Liouville-integrable systems exploit the Hamiltonian in the Euler–Lagrange equations of motion and the solution is expressible in terms of first integrals or equivalent integral equations.

The original contribution of this chapter is the derivation of a global first-order first integral which is achieved by considering the concept of Liouville integrability for a particular class of post-Newtonian dynamical systems. The derivation is simplified by employing the energy integral (6.40) which is discussed in Chap. 4.

The chapter is organised as follows: in Sects. 6.4 and 6.5 we demonstrate the integrability of classical separable and Liouville dynamical systems by following the detailed treatment of Whittaker (1988) and Ó Mathúna (2008). In Sects. 6.6 and 6.7, we discuss the classical Euler problem and derive the corresponding first integrals as an example of Sect. 6.5. In Sect. 6.8 we consider the concept of Liouville integrability for post-Newtonian gravitational systems and derive a new global constant of the motion in terms of generalised coordinates. Finally, in Sect. 6.9 we give an explicit illustrative example arising from the relativistic two-centre Euler problem in planar elliptical coordinates which extends the classical Newtonian results of Ó Mathúna (2008) (see Sect. 3.4). Finally, we note the work presented in this chapter is ongoing.

J. O'Leary, *General Relativistic and Post-Newtonian Dynamics for Near-Earth Objects and Solar System Bodies*, Springer Theses,
https://doi.org/10.1007/978-3-030-80185-4_6

6.2 Nomenclature

m_1, m_2	are the masses of the attracting central bodies
a	is the separation distance between the attracting bodies
U	is the gravitational potential function
μ_1, μ_2	are the gravitational parameters
G	is the Newtonian gravitational constant
V	is the classical gravitational potential energy
T	is the classical kinetic energy
$\mathcal{L}$	is the classical Lagrangian function
$\mathcal{E}$	is the total classical system energy
$\mathcal{H}$	is the classical Hamiltonian function
ξ, η	are the planar elliptical coordinates
$g_{\mu\nu}$	are the metric tensor components
c	is the speed of light in vacuum
U_j	are the vector gravitational potentials
δ_{ij}	is the Kronecker delta function
$\mathcal{L}$	is the post-Newtonian Lagrangian function
$\mathcal{H}$	is the post-Newtonian Hamiltonian function.

6.3 Background and Motivation

The planar Newtonian two-centre problem is a dynamical system of particular interest due to its direct integrability first discovered by Euler which was later generalised by Jacobi for the three-dimensional case. The two-centre problem is a special case of the well-known three body problem (Musielak and Quarles 2014) and according to Ó Mathúna (2008), it is one of only three fully integrable problems in celestial mechanics; the remaining two being the well-known Kepler and Vinti problems (see Vinti 1959 for example).

The gravitational two-centre problem (sometimes referred to as the *Euler problem*) describes the motion of a test particle in the gravitational field of two fixed masses m_1 and m_2 which are assumed to be non-interacting. We assume the attracting bodies equidistant from the origin and have positions $(0, -a)$ and $(0, a)$ in the (x, z) plane respectively (see Fig. 6.1 for geometry of problem). As the separation distance $2a$ between the non-interacting fixed centres tends to 0, we recover the classical Kepler problem as a degenerate case.

The *gravitational potential* U per unit mass at P is then given by

$$U = \frac{\mu_1}{\sqrt{x^2 + (z + a)^2}} + \frac{\mu_2}{\sqrt{x^2 + (z - a)^2}}, \tag{6.1}$$

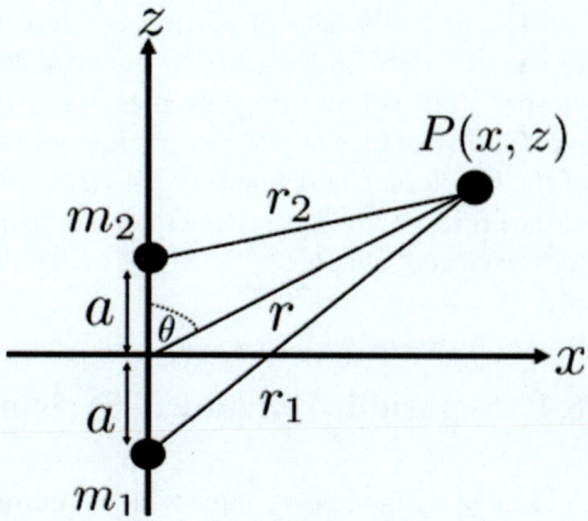

Fig. 6.1 Geometry of gravitational two-centre problem

where μ_1, μ_2 are constants proportional to the mass of each attracting body often referred to as *gravitational parameters* and are defined as $\mu_i = Gm_i$ where G denotes the classical Newtonian gravitational constant. The *gravitational potential energy* V is defined as the negative of the potential function such that $V \equiv -U$.

The Euler problem is over 200 years old and has appeared in the literature in many variations. The general relativistic analogue of the two or n-centre problem is described by the family of exact solutions of the Einstein–Maxwell vacuum field equations derived independently by Majumdar (1947) and Papapetrou (1947).

Hartle and Hawking (1972) provided the formal analytical extension of the *Majumdar-Papapetrou* framework and showed the solutions can be interpreted as a system of charged Reissner-Nordström black holes (see Chandrasekhar and Thorne 1985 and d'Inverno 1992 for a discussion on charged black holes) in equilibrium due to the balancing of electrostatic repulsion and gravitational attraction (Hartle and Hawking 1972).

Will (2009) derives a Carter-like constant associated with a test particle orbiting an axisymmetric central body for classical Newtonian theory and attempts to generalise this result for the general relativistic Euler problem, which the author refers to as the Bach-Weyl solution, and is discussed in Mirshekari and Will (2010). Mirshekari and Will (2010) show the general relativistic and first post-Newtonian Euler problems do not posses a corresponding Carter-like constant. It was later noted by Kopeikin (2018), the newly obtained Carter-like constant derived in Will (2009) is equivalent to the well-known Vinti integral in classical celestial mechanics given in an alternative form.

The solution procedure for the classical Euler problem appears briefly in Wintner (1941) and Whittaker (1988) with the complete analytical treatment given in Ó Mathúna (2008). Ó Mathúna (2008) gives a highly detailed account of the historical and theoretical development of the Euler problem and provides an extensive bibliography on the classical and quantum mechanical developments to which the reader is referred.

More recently, and of particular interest to the present chapter, is the work of Kobayashi and Yamada (2019) who provide a detailed analysis and characterisation of spacecraft orbits using the mass ratio of the fixed bodies and their respective separation distance as parameters. It would appear, given the theoretical significance of the Newtonian and general relativistic two-centre problems that the development of the corresponding post-Newtonian dynamics be worthwhile and is the motivation for the current chapter.

6.4 Separable Dynamical Systems

A class of dynamical systems which are immediately integrable are those classified as separable (Whittaker 1988; Ó Mathúna 2008). Separable systems with n degrees of freedom have kinetic and potential energy, T and V respectively, according to

$$T = \frac{1}{2}\sum_{j=1}^{n} f_j(q_j)\dot{q}_j^2 , \qquad V = \sum_{j=1}^{n} h_j(q_j), \tag{6.2}$$

where $f_j(q_j)$ and $h_j(q_j)$ denote arbitrary, scalar functions of the single generalised coordinate q_j. The kinetic and potential energies of separable systems involve the sum of n components respectively, where each jth component depends only on the jth generalised coordinate q_j. For example, for $n = 3$, the kinetic and potential energies (6.2) are given by

$$T = \frac{1}{2}\left(f_1(q_1)\dot{q}_1^2 + f_2(q_2)\dot{q}_2^2 + f_3(q_3)\dot{q}_3^2\right), \qquad V = h_1(q_1) + h_2(q_2) + h_3(q_3), \tag{6.3}$$

and total time derivatives are denoted using the standard *dot* notation. The n associated energy functions of (6.2) give rise to the solution of the dynamical system in terms of quadratures, demonstrated as follows:

The Lagrangian $\mathcal{L}$ associated with (6.2) is given by

$$\mathcal{L} \equiv T - V = \sum_{j=1}^{n}\left(\frac{1}{2} f_j(q_j)\,\dot{q}_j^2 - h_j(q_j)\right),$$

so that the Euler–Lagrange equation (Goldstein 1965) associated with the ith coordinate is given by

$$\frac{d}{dt}\left(f_i\dot{q}_i\right) - \frac{\dot{q}_i^2}{2}\frac{\partial f_i}{\partial q_i} = -\frac{\partial h_i}{\partial q_i}, \tag{6.4}$$

which, is equivalently expressed as

$$f_i \ddot{q}_i + \frac{1}{2}\frac{\partial f_i}{\partial q_i}\dot{q}_i^2 + \frac{\partial h_i}{\partial q_i} = 0. \tag{6.5}$$

Now since f_i and h_i depend only on the corresponding generalised coordinate, the total time derivatives are simply $df_i/dt = \dot{q}_j \partial f_i/\partial q_j$, with a similar expression for dh_i/dt.

Hence, the equation produced following multiplication of (6.5) by $\dot{q}_i$ is immediately integrable and is given by

$$\frac{1}{2} f_i(q_i)\, \dot{q}_i^2 + h_i(q_i) = c_i, \tag{6.6}$$

where the c_i denote n arbitrary constants of integration. We may identify (6.6) as the ith energy function for which the constants of integration are subject to the constraint (Ó Mathúna 2008)

$$\sum_{i=1}^{n} c_i = \mathcal{E},$$

where $\mathcal{E}$ is a constant denoting the total energy of the system. All dynamical systems discussed in this chapter are explicitly time independent, so that the Hamiltonian $\mathcal{H}$ and total system energy $\mathcal{E}$ are equivalent (Goldstein 1965).

A further n constants of integration γ_i are obtained by expressing Eq. (6.6) in the equivalent integral form

$$\int \mathrm{d}t = \gamma_i + \int \left[\frac{f_i(q_i)}{2\,(c_i - h_i(q_i))}\right]^{1/2} \mathrm{d}q_i, \tag{6.7}$$

and (6.7) constitutes the explicit solution of the system (Whittaker 1988). Equation (6.7) is further simplified by introducing new coordinates q_i^* according to the differential relations

$$\mathrm{d}q_i^* = \sqrt{f_i(q_i)}\mathrm{d}q_i, \qquad \dot{q}_i^* = \sqrt{f_i(q_i)}\dot{q}_i, \tag{6.8}$$

and is given by

$$\int \mathrm{d}t = \gamma_i + \int \frac{\mathrm{d}q_i^*}{\left[2\left(c_i - h_i^*(q_i^*)\right)\right]^{1/2}}, \tag{6.9}$$

where we have re-labelled the arbitrary scalar functions h_i according to $h_i(q_i) = h_i^*(q_i^*)$. Hence, and without loss of generality, the kinetic and potential energy for separable systems (6.2) can always be expressed as

$$T = \frac{1}{2}\sum_{j=1}^{n} (\dot{q}_j^*)^2, \qquad V = \sum_{j=1}^{n} h_j^*(q_j^*). \tag{6.10}$$

6.5 Liouville-Integrable Systems

A further class of dynamical systems which are integrable are those prescribed with kinetic and potential energies (Whittaker 1988)

$$T = \frac{1}{2}\mathcal{G}\sum_{j=1}^{n} f_j(q_j)\dot{q}_j^2, \qquad V = \mathcal{G}^{-1}\sum_{j=1}^{n} h_j(q_j), \tag{6.11}$$

where $\mathcal{G}$ is given by

$$\mathcal{G}(q_k) = \sum_{k=1}^{n} g_k(q_k), \tag{6.12}$$

and $g_k(q_k)$ denote a further sum of n arbitrary scalar functions where the kth component is a function of only the kth generalised coordinate. For example, for $n = 3$, the kinetic and potential energies for systems of the form (6.11) are given by

$$\begin{aligned} T &= \frac{1}{2}\left(g_1(q_1) + g_2(q_2) + g_3(q_3)\right)\left(f_1(q_1)\dot{q}_1^2 + f_2(q_2)\dot{q}_2^2 + f_3(q_3)\dot{q}_3^2\right), \\ V &= \left(h_1(q_1) + h_2(q_2) + h_3(q_3)\right) / \left(g_1(q_1) + g_2(q_2) + g_3(q_3)\right), \end{aligned}$$

and systems of the form (6.11) are referred to as Liouville-integrable dynamical systems.

In order to reduce a system in the form (6.11) to n first integrals and separable quadratures respectively, we proceed as follows.

We now assume that we have made the coordinate transformation (6.8) for (6.11) and we have dropped the star, so that the Euler–Lagrange equation associated with the ith generalised coordinate is given by

$$\frac{\mathrm{d}}{\mathrm{d}t}\left(\mathcal{G}\dot{q}_i\right) - \frac{1}{2}\frac{\partial \mathcal{G}}{\partial q_i}\sum_{j=1}^{n}\dot{q}_j^2 = -\frac{\partial V}{\partial q_i}, \tag{6.13}$$

so that following multiplication by $2\mathcal{G}\dot{q}_i$ and substitution of the classical total system energy

$$\mathcal{H} = \mathcal{E} \equiv T + V, \tag{6.14}$$

Equation (6.13) becomes

$$\frac{\mathrm{d}}{\mathrm{d}t}\left(\mathcal{G}^2\dot{q}_i^2\right) = 2\dot{q}_i\frac{\partial}{\partial q_i}\left[(\mathcal{H} - V)\,\mathcal{G}\right]. \tag{6.15}$$

After some simple algebraic manipulation and using the definition for total time derivative of a separable function, Eq. (6.15) can be expressed in integrable form

$$\frac{\mathrm{d}}{\mathrm{d}t}\left(\mathcal{G}^2\dot{q}_i^2\right) = 2\frac{\mathrm{d}}{\mathrm{d}t}\left(\mathcal{H}g_i - h_i\right), \tag{6.16}$$

which is equivalently expressed as

$$\frac{1}{2}\left(\mathcal{G}^2\dot{q}_i^2\right) - \mathcal{H}g_i + h_i = C_i, \tag{6.17}$$

where here the n constants of integration are given by C_i. Finally, we may deduce a constraint on the sum of the constants of integration

$$\sum_{i=1}^{n} C_i = \mathcal{G}^2 \sum_{i=1}^{n} \frac{1}{2}\dot{q_i}^2 - \mathcal{H}\sum_{i=1}^{n} g_i + \sum_{i=1}^{n} h_i = 0, \tag{6.18}$$

and the remaining n constants of integration required to fully determine the solution of the Liouville system are obtained by expressing (6.17) in integral form.

6.6 The Classical Euler Problem

The classical Euler problem is evidently soluble following a coordinate transformation which leaves the system of Liouville type. The standard approach adopted in the literature is to introduce planar elliptical coordinates (ξ, η), which are sometimes referred to as planar prolate spheroidal coordinates (Ó Mathúna 2008) through the relations

$$x = a\sinh\xi\sin\eta, \qquad y = 0, \qquad z = a\cosh\xi\cos\eta, \tag{6.19}$$

where for constant ξ and η the expressions

$$\left(\frac{x}{a\sinh\xi}\right)^2 + \left(\frac{z}{a\cosh\xi}\right)^2 = 1, \qquad \left(\frac{x}{a\sin\eta}\right)^2 - \left(\frac{z}{a\cos\eta}\right)^2 = 1, \tag{6.20}$$

describe a particular family of elliptic and hyperbolic curves.

One of the underlying reasons for the integrability of the Euler problem is that *any* ellipse or hyperbola is an exact orbit when *either one* of the foci are attracting gravitational bodies. If both foci are attracting bodies, an exact orbit is also possible by superimposing the gravitational fields. This is a consequence of Bonnet's theorem which reads (taken verbatim from Whittaker (1988)):

Theorem 6.6.1 *If a given orbit can be described in each of n given fields of force, taken separately, the velocities at any point P of the orbit being* $v_1, v_2, \cdots, v_n$, *respectively, then the same orbit can be described in the field of force which is obtained by superimposing all these fields, the velocity at the point P being* $(v_1^2 + v_2^2 + \cdots + v_n^2)^{1/2}$.

The metric tensor components g_{mn} $(m, n = 1, 2)$ and the magnitude of the line element $(\mathrm{d}s)^2$ are given by

$$g_{mn} = a^2 \,\mathrm{diag}(\cosh^2 \xi - \cos^2 \eta, \cosh^2 \xi - \cos^2 \eta), \tag{6.21}$$

and

$$(\mathrm{d}s)^2 = a^2 \left(\cosh^2 \xi - \cos^2 \eta\right) \left(\mathrm{d}\xi^2 + \mathrm{d}\eta^2\right), \tag{6.22}$$

respectively and the kinetic and potential energies are given by

$$T = \frac{1}{2} g_{mn} \dot{x}^m \dot{x}^n = \frac{a^2}{2} \left(\cosh^2 \xi - \cos^2 \eta\right) \left(\dot{\xi}^2 + \dot{\eta}^2\right), \tag{6.23}$$

and

$$V = -\frac{\mu_1}{a\left(\cosh \xi + \cos \eta\right)} - \frac{\mu_2}{a\left(\cosh \xi - \cos \eta\right)}. \tag{6.24}$$

The gravitational potential can then be expressed in simplified form according to (Ó Mathúna 2008)

$$V = -\mathcal{M} \frac{a \cosh \xi + \beta a \cos \eta}{a^2 \cosh^2 \xi - a^2 \cos^2 \eta}, \tag{6.25}$$

where $\mathcal{M} = G\left(m_2 + m_1\right)$ and the dimensionless mass parameter $\beta \equiv (m_2 - m_1) / (m_2 + m_1)$.

The Lagrangian $\mathcal{L} = \mathcal{L}(\xi, \eta, \dot{\xi}, \dot{\eta})$ associated with (6.23) and (6.25) is then given by

$$\mathcal{L} = \frac{1}{2} \left[g_1(\xi) + g_2(\eta)\right] \left(\dot{\xi}^2 + \dot{\eta}^2\right) + \frac{h_1(\xi) + h_2(\eta)}{g_1(\xi) + g_2(\eta)}, \tag{6.26}$$

where we have introduced

$$g_1(\xi) = a^2 \cosh^2 \xi, \qquad g_2(\eta) = -a^2 \cos^2 \eta, \tag{6.27}$$

$$h_1(\xi) = -\mathcal{M} a \cosh \xi, \qquad h_2(\eta) = -\mathcal{M} \beta a \cos \eta, \tag{6.28}$$

and we immediately recognise the problem as integrable as inferred by Liouville.

6.7 First Integrals of Classical Euler Problem

The energy function $\mathcal{E} = \mathcal{H}$ is derived by comparing the sum of the Euler–Lagrange equations with the total time derivative of the Lagrangian and is given by

$$\dot{\xi} \frac{\partial \mathcal{L}}{\partial \dot{\xi}} + \dot{\eta} \frac{\partial \mathcal{L}}{\partial \dot{\eta}} - \mathcal{L} = \mathcal{H}. \tag{6.29}$$

The Euler–Lagrange equations for ξ and η in Liouville coordinates (6.27) are given by

$$\begin{aligned}\mathcal{G}\dot{\xi}\frac{\mathrm{d}}{\mathrm{d}t}\left(\mathcal{G}\dot{\xi}\right) - \frac{\mathcal{G}\dot{\xi}}{2}\frac{\mathrm{d}\mathcal{G}}{\mathrm{d}\xi}\left(\dot{\xi}^2+\dot{\eta}^2\right) &= -\mathcal{G}\dot{\xi}\frac{\partial V}{\partial \xi},\\ \mathcal{G}\dot{\eta}\frac{\mathrm{d}}{\mathrm{d}t}\left(\mathcal{G}\dot{\eta}\right) - \frac{\mathcal{G}\dot{\eta}}{2}\frac{\mathrm{d}\mathcal{G}}{\mathrm{d}\eta}\left(\dot{\xi}^2+\dot{\eta}^2\right) &= -\mathcal{G}\dot{\eta}\frac{\partial V}{\partial \eta},\end{aligned} \tag{6.30}$$

respectively, where we have multiplied each respective Euler–Lagrange equation by $\mathcal{G}\dot{\xi}$ and $\mathcal{G}\dot{\eta}$. Now using (6.14), we can express the ξ-Euler–Lagrange equation in equivalent integrable form according to

$$\begin{aligned}\frac{1}{2}\frac{\mathrm{d}}{\mathrm{d}t}\left(\mathcal{G}\dot{\xi}\right)^2 &= \dot{\xi}\left[\mathcal{H}\frac{\mathrm{d}g_1}{\mathrm{d}\xi} - \frac{\mathrm{d}}{\mathrm{d}\xi}\left(\mathcal{G}V\right)\right]\\ &= \frac{\mathrm{d}}{\mathrm{d}t}\left[\mathcal{H}g_1(\xi) - h_1(\xi)\right],\end{aligned} \tag{6.31}$$

and the first integral associated with ξ is given by

$$\frac{1}{2}\left(\mathcal{G}\dot{\xi}\right)^2 + h_1(\xi) - \mathcal{H}g_1(\xi) = \xi_0, \tag{6.32}$$

and by symmetry of the problem, the remaining first integral for η is given by

$$\frac{1}{2}\left(\mathcal{G}\dot{\eta}\right)^2 + h_2(\eta) - \mathcal{H}g_2(\eta) = \eta_0, \tag{6.33}$$

where ξ_0 and η_0 denote arbitrary constants of their respective first integral. By virtue of energy conservation, the constraint (6.18) is satisfied and is given by

$$\xi_0 + \eta_0 = \mathcal{G}\left[\frac{1}{2}\mathcal{G}\left(\dot{\xi}^2+\dot{\eta}^2\right) + \frac{h_1(\xi)+h_2(\eta)}{\mathcal{G}} - \mathcal{H}\right] = 0. \tag{6.34}$$

The first integrals (6.29), (6.32) and (6.33) constitute the solution to the classical Euler problem.

6.8 Post-Newtonian Liouville Systems

In this section, we consider post-Newtonian systems in the framework of Liouville. We adopt the same formal procedure outlined in Sect. 6.5 using the corresponding relativistic quantities. The critical step in the classical case is to exploit the energy integral in the Euler–Lagrange equation (6.15). The form of the energy integral used in this section has been derived in Chap. 4 and is given by (6.40).

We assume a static spacetime and adopt the metric tensor components $g_{\mu\nu}$ prescribed by the Damour, Soffel, Xu formalism (see Chap. 3). Again, Greek indices ($\mu, \nu = 0, 1, 2, 3$) account for spacetime variables, with the 0th component being reserved for time. Further, we assume the scalar and vector gravitational potentials take the form

$$U = -V = -\mathcal{G}^{-1} \sum_{j=1}^{n} h_j(q_j), \qquad U_j = 0, \tag{6.35}$$

respectively, where $\mathcal{G}$ is defined according to (6.12) and the assumption $(6.35)_b$ is a consequence of the static spacetime. It should be noted that since the first post-Newtonian approximation is truncated to $O(c^{-2})$, then any dynamical quantity such as velocity, acceleration or energy which are of $O(c^{-2})$ can be replaced, as a first order approximation, using the corresponding Newtonian counterpart. This is demonstrated explicitly in Eq. (6.40).

The post-Newtonian Lagrangian $\mathcal{L}$ is derived using a variational principle and is defined according to Eq. (4.15) and is given explicitly to $O(c^{-2})$ by

$$\mathcal{L} = \frac{\mathcal{G}}{2} \sum_{j=1}^{n} \dot{q}_j^2 + U - \frac{U^2}{2c^2} + \frac{3U}{2c^2} \mathcal{G} \sum_{j=1}^{n} \dot{q}_j^2 + \frac{1}{8c^2} \left(\mathcal{G} \sum_{j=1}^{n} \dot{q}_j^2 \right)^2 + O(c^{-4}), \tag{6.36}$$

and the post-Newtonian Hamiltonian $\mathcal{H}$ is derived using the Legendre transformation (Goldstein 1965)

$$\mathcal{H} = \dot{q}_j \frac{\partial \mathcal{L}}{\partial \dot{q}_j} - \mathcal{L}, \tag{6.37}$$

given explicitly by

$$\mathcal{H} = \frac{\mathcal{G}}{2} \sum_{j=1}^{n} \dot{q}_j^2 - U + \frac{U^2}{2c^2} + \frac{3U}{2c^2} \mathcal{G} \sum_{j=1}^{n} \dot{q}_j^2 + \frac{3}{8c^2} \left(\mathcal{G} \sum_{j=1}^{n} \dot{q}_j^2 \right)^2 + O(c^{-4}), \tag{6.38}$$

where we have assumed the magnitude of the post-Newtonian coordinate velocity v^2 is of the form $v^2 = \mathcal{G} \sum_{j=1}^{n} \dot{q}_j^2$. In order to obtain a simplified post-Newtonian expression for the magnitude of the coordinate velocity as given in Sect. 6.5, we make the substitution

$$\mathcal{G} \sum_{j=1}^{n} \dot{q}_j^2 = 2 \left(\mathcal{H} + U \right), \tag{6.39}$$

in the velocity terms proportional to c^{-2} in Eq. (6.38) which yields

$$\frac{\mathcal{G}}{2} \sum_{j=1}^{n} \dot{q}_j^2 = \mathcal{H} \left(1 - \frac{6U}{c^2} \right) + U - \frac{5U^2}{c^2}, \tag{6.40}$$

where we have ignored terms of order $O(\mathcal{H}^2/c^2)$ as suggested by Poisson and Will (2014) and is discussed further at the end of this section.

The ith Euler–Lagrange equation associated with the post-Newtonian Lagrangian (6.36) is given by

$$\frac{\mathrm{d}}{\mathrm{d}t}\left(\mathcal{G}\dot{q}_i + \frac{3\mathcal{G}U}{c^2}\dot{q}_i + \frac{\mathcal{G}^2\dot{q}_i}{2c^2}\sum_{m=1}^{n}\dot{q}_m^2\right) = \frac{1}{2}\frac{\partial\mathcal{G}}{\partial q_i}\sum_{m=1}^{n}\dot{q}_m^2 + \frac{\partial U}{\partial q_i} - \frac{1}{c^2}\left[U\frac{\partial U}{\partial q_i} - \frac{3}{2}\left(\mathcal{G}\frac{\partial U}{\partial q_i} + U\frac{\partial\mathcal{G}}{\partial q_i}\right)\sum_{m=1}^{n}\dot{q}_m^2 - \frac{\mathcal{G}}{4}\frac{\partial\mathcal{G}}{\partial q_i}\left(\sum_{m=1}^{n}\dot{q}_m^2\right)^2\right], \tag{6.41}$$

which is further simplified following multiplication of (6.41) by $\mathcal{G}\dot{q}^i$ and substitution of (6.40) which is given by

$$\frac{1}{2}\left(1+\frac{\mathcal{H}}{c^2}\right)\frac{\mathrm{d}}{\mathrm{d}t}\left(\mathcal{G}^2\dot{q}_i^2\right) + \frac{4\mathcal{G}\dot{q}_i}{c^2}\frac{\mathrm{d}}{\mathrm{d}t}\left(U\mathcal{G}\dot{q}_i\right) = \mathcal{D}_i\left[\mathcal{G}\left(\mathcal{H}+U\right)\right] - \frac{1}{c^2}\left[U\left(\mathcal{H}+U\right)\mathcal{D}_i\mathcal{G} - \mathcal{G}\left(3\mathcal{H}+2U\right)\mathcal{D}_iU\right], \tag{6.42}$$

where we have introduced the differential operator $D_i \equiv \dot{q}_i\partial/\partial q_i$ and it is clear that in the Newtonian limit ($c \to \infty$) we recover the classical Liouville result (6.16). At this point, we note that it is not obvious whether Eq. (6.42) can be expressed in integrable form for each individual Euler–Lagrange equation and further work is required.

However, we may express (6.42) as a global first integral by performing a summation over the index i and is given by

$$\frac{1}{2}\left(1+\frac{\mathcal{H}}{c^2}\right)\frac{\mathrm{d}}{\mathrm{d}t}\left(\mathcal{G}^2\sum_{i=1}^{n}\dot{q}_i^2\right) = \frac{\mathrm{d}}{\mathrm{d}t}\left\{\left[1-\frac{5U}{c^2}\right]\left[\mathcal{G}\left(\mathcal{H}+U\right)\right]\right\}, \tag{6.43}$$

and Eq. (6.43) integrates to give

$$\frac{1}{2}\left(1+\frac{\mathcal{H}}{c^2}\right)\left(\mathcal{G}^2\sum_{i=1}^{n}\dot{q}_i^2\right) - \left(1-\frac{5U}{c^2}\right)\left[\mathcal{G}\left(\mathcal{H}+U\right)\right] = \mathcal{C}, \tag{6.44}$$

where $\mathcal{C}$ denote a global constant of integration. Further, it can be shown, up to first post-Newtonian order, the constant of integration satisfies the constraint $\mathcal{C} \equiv 0$ which is a consequence of energy conservation. The constraint of the global first integral can be demonstrated in the following two equivalent ways. First, we express Eq. (6.44) in equivalent form according to

$$\mathcal{G}\left[\frac{1}{2}\left(1+\frac{\mathcal{H}}{c^2}\right)\mathcal{G}\sum_{i=1}^{n}\dot{q}_i^2-(\mathcal{H}+U)+\frac{5\mathcal{H}U}{c^2}+\frac{5U^2}{c^2}\right]=\mathcal{C}. \quad (6.45)$$

Following this, and using Eq. (6.40), we substitute the expression for $5U^2/c^2$ in (6.45), which only leaves terms of order $O(\mathcal{H}^2/c^2)$.

We note Poisson and Will (2014) state that due to the ambiguity and non-uniqueness associated with the post-Newtonian first integral for energy, we can always add an arbitrary amount of $\mathcal{H}^2/c^2$ to Eq. (6.40). Hence, we can ignore the additional term of order $O(\mathcal{H}^2/c^2)$ above and deduce $\mathcal{C}=0$. Alternatively, we can show $\mathcal{C}=0$ by adding the requirement that in the Newtonian limit Eq. (6.45) reduces to Eq. (6.18).

6.9 Post-Newtonian Euler Problem

In this section, we consider the post-Newtonian two-centre problem as an illustration of the post-Newtonian Liouville formalism and we derive the corresponding first integrals using (6.44). The geometry of the post-Newtonian Euler problem is equivalent to the Newtonian case and is described in Fig. 6.1.

The magnitude of the post-Newtonian line element and coordinate velocity in terms of planar elliptical coordinates (6.19) are given by

$$(\mathrm{d}s)^2=\left(1-\frac{2U}{c^2}+\frac{2U^2}{c^4}\right)c^2\mathrm{d}t^2-a^2\left(1+\frac{2U}{c^2}\right)\left(\cosh^2\xi-\cos^2\eta\right)\left(\mathrm{d}\xi^2+\mathrm{d}\eta^2\right), \quad (6.46)$$

and

$$v^2=\mathcal{G}\sum_{i=1}^{n}\dot{q}_j^2=a^2\left(\cosh^2\xi-\cos^2\eta\right)\left(\dot{\xi}^2+\dot{\eta}^2\right), \quad (6.47)$$

respectively where $\mathcal{G}$ is defined explicitly as

$$\mathcal{G}=a^2\left(\cosh^2\xi-\cos^2\eta\right). \quad (6.48)$$

Finally, we define U according to

$$U=\mathcal{M}\frac{a\cosh\xi+\beta a\cos\eta}{a^2\cosh^2\xi-a^2\cos^2\eta}, \quad (6.49)$$

where $\mathcal{M}$ and β are defined in Sect. 6.6. According to (6.44) the global post-Newtonian first integral for the relativistic Euler problem is given by

$$\frac{1}{2}\left(1+\frac{\mathcal{H}}{c^2}\right)\mathcal{G}^2\left(\dot{\xi}^2+\dot{\eta}^2\right)-\mathcal{H}\left(g_1+g_2\right)-\left(h_1+h_2\right)\left(1-\frac{5\mathcal{H}}{c^2}-\frac{5U}{c^2}\right)=0 \quad (6.50)$$

where $\mathcal{H}$ is a constant, $\mathcal{G}$ and U are given explicitly by (6.48) and (6.49) respectively and we have introduced

$$\begin{aligned} h_1(\xi) &= \mathcal{M}a\cosh\xi, \qquad h_2(\eta) = \mathcal{M}\beta a\cos\eta, \\ g_1(\xi) &= a^2\cosh^2\xi, \qquad g_2(\eta) = -a^2\cos^2\eta. \end{aligned} \tag{6.51}$$

We note that Eq. (6.50) does not appear to have been given previously in the literature. As a concluding remark, we note Eq. (6.50) may be re-arranged to give

$$\begin{aligned} &\left(\cosh^2\xi - \cos^2\eta\right)^2\left(\dot{\xi}^2 + \dot{\eta}^2\right) = \\ &\qquad \alpha\left(\cosh^2\xi - \cos^2\eta\right) + \gamma\left(\cosh\xi + \beta\cos\eta\right)\left(1 - \frac{5\mathcal{H}}{c^2} - \frac{5U}{c^2}\right), \end{aligned} \tag{6.52}$$

where α and γ denote two new constants defined by

$$\alpha \equiv \frac{2\mathcal{H}}{a^2\left(1 + \mathcal{H}/c^2\right)}, \quad \gamma \equiv \frac{2\mathcal{M}}{a^3\left(1 + \mathcal{H}/c^2\right)}, \tag{6.53}$$

where it is clear that further progress may be made only through numerical procedures. As a check on our calculations, we note that in the Newtonian limit ($c \to \infty$) we recover the classical results given in Whittaker (1988) and Ó Mathúna (2008), and on addition of Eqs. (6.52), we obtain

$$\frac{v^2}{2} = \frac{\mathcal{H} + U}{1 + \mathcal{H}/c^2}\left(1 - \frac{5U}{c^2}\right), \tag{6.54}$$

and this is equivalent to the energy integral (6.40) used at the outset.

6.10 Conclusions and Future Work

In this work, we have considered the concept of Liouville integrability for first post-Newtonian systems. We derived a corresponding global first integral (6.52) associated with the post-Newtonian Euler problem by adopting the same formal procedure discussed in Wintner (1941), Whittaker (1988) and Ó Mathúna (2008). A major difference between the resulting formalism is the post-Newtonian first integrals are not readily separable. This is due to the appearance of non-linear gravitational potential terms in the Lagrangian and Hamiltonian which is in good agreement with Drake et al. (1996) who studied the chaotic motion of charged particles for the special relativistic Euler-problem.

Although the contents of the present chapter are comprehensive, further work is possible. For example, Drake et al. (1996) has studied the chaotic dynamics of a charged test particle orbiting two fixed, charged centres in flat Minkowski space and Contopoulos (1990) has studied the general relativistic analogue. It is of interest to the present author to investigate the corresponding dynamics in the post-Newtonian framework and determine the various orbit types. Further, very little research exists in the field of post-Newtonian satellite orbits for two-centre problems. Ó Mathúna (2008) provides the complete analytical treatment of the classical Euler problem for the planar and three-dimensional problems where the generalisation to post-Newtonian dynamics is yet to be complete.

Part IV
General Relativistic Orbits

Chapter 7
General Relativistic Orbit Determination

7.1 Introduction

The classical determination of orbits (see for example Vallado 2001; Montenbruck and Gill 2012) for spacecraft and interplanetary probes relies on the numerical integration of Newton's second law of motion, combined with linearly superimposed perturbative forces. Specifically, this corresponds to the perturbed three-dimensional acceleration of the spatial coordinates for a given body with respect to the absolute coordinate time of Newtonian mechanics. This is the standard procedure followed by modern orbit determination packages, such as the French space agency GINS software (see Marty 2013 for technical documentation). Due to the rapid increase in measurement technology, relativistic effects need to be accounted for at some level. Examples include: improvements in tracking precision, such as interplanetary lasers; development and incorporation of atomic clocks on-board spacecraft and probes; and advances in geodetic techniques (Müller et al. 2008). Such relativistic effects (see Chap. 4 for a discussion) are analytically derived at the first post-Newtonian level, and incorporated in the three-dimensional perturbed acceleration as effective forces.

Pireaux et al. (2006) argues the classical orbit determination process is now reaching its limits in terms of complexity, and is further prone to errors when incorporating general relativity. The authors further suggest that the full general theory of relativity should form the basis of any future orbit determination software, which naturally embodies the classical and relativistic gravitational effects and any coupling between them (Pireaux et al. 2008). In this brief chapter, we outline a framework to calculate general relativistic orbits using a *generalised geodesic equation* (Lichnerowicz and Teichmann 1955; Charon 1963; Pireaux et al. 2006) in conjunction with data obtained from spacecraft and probes equipped with on-board accelerometers. We note this work is the formal extension of Pireaux et al. (2006) and is ongoing. A manuscript is currently under review titled:

J. O'Leary, *General Relativistic and Post-Newtonian Dynamics for Near-Earth Objects and Solar System Bodies*, Springer Theses,
https://doi.org/10.1007/978-3-030-80185-4_7

- O'Leary, J., Barriot, J.P., 2021. An application of symplectic integration for general relativistic planetary orbitography subject to non-gravitational forces.

7.2 Perturbed Equations of Motion

The motion of spacecraft or interplanetary probes, subject to no external perturbations, follow geodesics. However, such motion is ideal and not realistic. In reality, orbits are more complex which are subject to gravitational and non-gravitational perturbations (see Table 7.1) which need to be considered for the accurate determination of orbits. In classical Newtonian gravity, this is easily achieved by numerically evaluating the perturbed Kepler problem.

The framework suggested by Pireaux et al. (2006) takes advantage of data provided by on-board accelerometers combined with numerically integrating the following equation

$$\frac{\mathrm{d}u^\alpha}{\mathrm{d}\tau} = -\Gamma^\alpha_{\beta\gamma} u^\beta u^\gamma + K_\beta \left(g^{\alpha\beta} - \frac{u^\alpha}{c}\frac{u^\beta}{c} \right), \tag{7.1}$$

where the four-velocity, proper time of the spacecraft, Christoffel symbols of the second kind, speed of light in vacuum and inverse metric tensor components are given by u^α, τ, $\Gamma^\alpha_{\beta\gamma}$, c and $g^{\alpha\beta}$ respectively. The quantity K_β is a four-vector which contains components of the non-gravitational perturbations. For $K_\beta = 0$, we recover the standard geodesic equation of motion. We note the derivation of Eq. (7.1) has been relegated to the manuscript mentioned above. Alternatively, the reader is referred to Lichnerowicz and Teichmann (1955) and Charon (1963) for explicit derivations.

A distinct advantage of this approach emerges as a consequence of the field equations of gravitation. For example, gravitational phenomena are direct manifestations of spacetime geometry. The metric tensor components (and quantities derived from them) embody all information required to account for the gravitational perturba-

Table 7.1 Gravitational (G) and non-gravitational (NG) forces (Pireaux et al. 2006)

Perturbation (m/s^2)	LAGEOS	CHAMP
G: Monopole moment of the Earth	2.8	8.6
G: Low order sectoral harmonics ($l = m = 2$)	6.0×10^{-6}	6.4×10^{-5}
G: Sum of general relativistic contributions	9.5×10^{-10}	1.7×10^{-8}
NG: Atmospheric drag	3×10^{-12}	3.5×10^{-7}
NG: Solar radiation pressure	3.2×10^{-9}	3.2×10^{-8}
NG: Earth albedo pressure	3.4×10^{-12}	3.3×10^{-9}

tions acting on spacecraft and probes. Further, the non-gravitational four-vectors are derived directly from the energy-momentum tensor. In other words, the field equations which relate spacetime geometry with energy-momentum provide a direct connection between both contributions to the mechanics of spacecraft i.e. gravitational and non-gravitational.

Two further distinct advantages in using general relativity as the basis for future orbit determination software is the existence of a first integral in the local frame of the spacecraft, namely, the constancy of the magnitude of the four-velocity given by

$$\bar{g}_{\alpha\beta}\bar{u}^{\alpha}\bar{u}^{\beta} = \pm c^2, \tag{7.2}$$

where barred expressions denote quantities in the local frame and the sign of the constant is dependent on the metric signature.

In the local frame, the only *forces* acting on the spacecraft are those encoded in K_β. Hence, if the spatial components K_i of the four-vector are provided by on-board accelerometers, then the temporal component is obtained by differentiating (7.2) with respect to τ and solving for K_0 accordingly. Further, (7.2) provides a direct check on numerical calculations. For example, one can automatically verify any modifications or additions to the software by verifying that Eq. (7.2) remains constant in the local frame of the spacecraft. The quantity (7.2) does not exist in classical Newtonian orbit determination.

7.3 Numerical Integration Scheme

Due to the existence of Eq. (7.2), we choose to numerically integrate the equations of motion (7.1) coupled with (7.2) using a symplectic integration scheme. Butcher (1964) derives the coefficients required to construct a 5-stage, 10th order implicit Runge–Kutta scheme which satisfy the symplecticity constraint (Cooper 1987) given by

$$b_i a_{ij} + b_j a_{ji} - b_i b_j = 0, \qquad (i, j = 1, 2, \cdots s), \tag{7.3}$$

where s denotes the number of stages and the b_i, a_{ij} correspond to the components of a Butcher Tableau (Butcher 2016). The general form of a Butcher tableau is given by

$$\begin{array}{c|cccc} c_1 & a_{11} & a_{12} & \dots & a_{1s} \\ c_2 & a_{21} & a_{22} & \dots & a_{2s} \\ \vdots & \vdots & & & \vdots \\ c_s & a_{s1} & a_{s2} & \dots & a_{ss} \\ \hline & b_1 & b_2 & \dots & b_s \end{array}$$

where the c_i are defined according to

$$c_i \equiv \sum_{j=1}^{s} a_{ij}.$$

The explicit coefficients for the integration scheme are given by

$$\begin{array}{c|ccccc}
1/2-\omega_2 & \omega_1 & \omega_1'-\omega_3+\omega_4' & 32/225-\omega_5 & \omega_1'-\omega_3-\omega_4' & \omega_1-\omega_6 \\
1/2-\omega_2' & \omega_1-\omega_3'+\omega_4 & \omega_1' & 32/225-\omega_5' & \omega_1'-\omega_6' & \omega_1-\omega_3'-\omega_4 \\
1/2 & \omega_1+\omega_7 & \omega_1'+\omega_7' & 32/225 & \omega_1'-\omega_7' & \omega_1-\omega_7 \\
1/2+\omega_2' & \omega_1+\omega_3'+\omega_4 & \omega_1'+\omega_6' & 32/225+\omega_5' & \omega_1' & \omega_1+\omega_3'-\omega_4 \\
1/2+\omega_2 & \omega_1+\omega_6 & \omega_1'+\omega_3+\omega_4' & 32/225+\omega_5 & \omega_1'+\omega_3-\omega_4' & \omega_1 \\
\hline
 & 2\omega_1 & 2\omega_1' & 64/225 & 2\omega_1' & 2\omega_1
\end{array}$$

where

$$\omega_1 = \frac{322-13\sqrt{70}}{3600}, \qquad \omega_1' = \frac{322+13\sqrt{70}}{3600},$$

$$\omega_2 = \frac{1}{2}\sqrt{\frac{35+2\sqrt{70}}{63}}, \qquad \omega_2' = \frac{1}{2}\sqrt{\frac{35-2\sqrt{70}}{63}},$$

$$\omega_3 = \omega_2\left(\frac{452+59\sqrt{70}}{3240}\right), \qquad \omega_3' = \omega_2'\left(\frac{452-59\sqrt{70}}{3240}\right),$$

$$\omega_4 = \omega_2\left(\frac{64+11\sqrt{70}}{1080}\right), \qquad \omega_4' = \omega_2'\left(\frac{64-11\sqrt{70}}{1080}\right),$$

$$\omega_5 = 8\omega_2\left(\frac{23-\sqrt{70}}{405}\right), \qquad \omega_5' = 8\omega_2'\left(\frac{23+\sqrt{70}}{405}\right),$$

$$\omega_6 = \omega_2 - 2\omega_3 - \omega_5, \qquad \omega_6' = \omega_2' - 2\omega_3' - \omega_5',$$

$$\omega_7 = \omega_2\left(\frac{308-23\sqrt{70}}{960}\right), \qquad \omega_7' = \omega_2'\left(\frac{308+23\sqrt{70}}{960}\right).$$

In general, symplectic integration schemes for classical orbit determination are not as effective as standard schemes because quantities such as energy and angular momentum are no longer conserved when non-gravitational forces are considered. The existence of (7.2) in general relativity allows one to take advantage of structure-preserving integration schemes, which have been shown to admit a wide range of attractive properties when simulating complicated dynamical systems (McLachlan and Quispel 2006; Hairer et al. 2006).

7.4 Conclusions and Future Work

We have successfully developed a working prototype software in Fortran 90 using 128-bit quad precision formally extending the work of Pireaux et al. (2006) titled General Relativistic Accelerometer-based Propagation Environment (GRAPE). As a test case, we numerically integrate the perturbed equations of motion (7.1) for the NASA Parker solar probe,[1] in the non-rotating Schwarzschild field of the Sun. The software allows one to easily transform between local and global frames using the standard tetrad formalism. The explicit transformations can be found in Müller and Grave (2009). The accuracy of the software has been verified by implementing the integration scheme presented above.

As an extension to the work presented above, it is beneficial to model the motion of the test particle around a non-spherical body. In general relativity, this is achieved using the weak field approximation to the Schwarzschild metric (Drake 2006). Further, it is proposed that comparisons be made with the French space agency orbit determination software GINS. Finally, it is of interest to the present author to investigate the motion of a test particle, subject to non-gravitational forces, around a rotating body using the Kerr metric.

---

[1] See https://solarsystem.nasa.gov/missions/parker-solar-probe/in-depth/ for mission details.

Part V
Conclusion

Chapter 8
Concluding Remarks

8.1 Summary

In this thesis, we have employed applied mathematical modelling techniques to the field of general relativistic and post-Newtonian dynamics. The topics studied can be categorised into three broad areas:

- Exact solutions to the Einstein field equations of gravitation.
- Post-Newtonian dynamics and celestial mechanics.
- General relativistic orbit determination including non-gravitational perturbations.

8.2 Exact Solutions to the Field Equations of Gravitation

In Chap. 2, we derive an alternative spherically symmetric solution to the Einstein field equations of gravitation. The solution is obtained by assuming a non-orthogonal line element and the corresponding unknown metric tensor components are determined accordingly. The solution arises as one of three distinct possibilities with the remaining two situations giving rise to Schwarzschild-like solutions.

The so-called Birkhoff's theorem largely isolates the study of spherically symmetric vacuum solutions of the Einstein field equations and restricts attention to the Schwarzschild solution. However, once derived, coordinate transformations are required to remove spacetime singularities and achieve the maximal analytically extended spacetime. The motivation and goal of this chapter was to provide an alternative exact solution to the Einstein field equations which does not require further coordinate transformations to remove coordinate singularities. Finally, the author conjectures the same approach can be applied to other spherically symmetric solutions with non-zero source terms such as the well-known Reissner–Nordström solution which also requires further transformations following derivation.

J. O'Leary, *General Relativistic and Post-Newtonian Dynamics for Near-Earth Objects and Solar System Bodies*, Springer Theses,
https://doi.org/10.1007/978-3-030-80185-4_8

8.3 Post-Newtonian Dynamics and Celestial Mechanics

Chapter 3 contains an extensive review of the early development of the post-Newtonian approximation and is intended to serve as a stand-alone introduction to the Damour, Soffel, Xu post-Newtonian formalism for relativistic celestial mechanics.

In Chap. 4, we present a novel derivation of the first post-Newtonian energy integral in the local geocentric frame. This is achieved by seeking a Jacobi-like integral associated with the first post-Newtonian equations of motion. The integral so obtained contains an exponential function and when linearly approximated can be shown to coincide with expressions found in the literature. In order to compare the newly derived integral with those in the literature, we perform an extensive numerical investigation into the long-term evolution of the energy integrals using high order symplectic integration schemes. We find the long-term error in the expressions found in the literature is significantly greater in magnitude. This suggests retaining the exponential function may be beneficial for post-Newtonian celestial mechanics.

Chapter 5 is concerned with the post-Newtonian two-body problem. After deriving the two-body equations of motion in the centre of mass frame, we present the derivation of the post-Newtonian Binet equation. This is achieved by exploiting the conservation of angular momentum in the radial equation of motion. Following this, we derive a new second order, ordinary differential equation which characterises the shape of post-Newtonian satellite orbits. The resulting differential equation contains multiple exponential functions of different orders. When comparing the post-Newtonian Binet equation and the corresponding linear approximation with the full general relativistic Binet equation, we deduce that retaining the exponential function is advantageous and closely resembles the orbits of general relativity.

Finally, in Chap. 6 we consider the concept of Liouville integrability to a class of post-Newtonian dynamical systems. We follow the same formal procedure as the classical Newtonian case and derive a first integral for an arbitrary gravitational potential. As an illustrative example, we consider the post-Newtonian two-centre Euler problem and derive the corresponding first integral in prolate spheroidal coordinates. We note the resulting formalism is not integrable as is the case for the classical Newtonian counterpart. This is due to the appearance of a non-linear potential term and requires further investigation.

8.4 General Relativistic Orbit Determination

In Chap. 7, we provide an overview of a new, high precision, general relativistic orbit determination tool. We suggest a novel framework to numerically integrate the Schwarzschild geodesic equations of motion which includes non-gravitational perturbations. The framework encodes the non-gravitational forces in a four-vector which enter the right-hand side of the standard geodesic equation. By utilising high-precision data from on-board accelerometers, we suggest a method to determine the spatial components of non-gravitational perturbations. The temporal component is then obtained by exploiting the constancy of the four-velocity in the local tetrad. The final contribution of the method outlined above is to exploit the structure-preserving properties of geometric numerical integrators, which provide a reliable framework to verify the accuracy of the new scheme. This is achieved by ensuring the norm of the four-velocity remains constant throughout simulations. We note, it is impractical to suggest that the newly developed software may serve as an active replacement for current orbit propagation tools such as the French space agency GINS. However, the software can serve as a general relativistic test-bed for current and future orbit determination software.

Appendix A

A.1 Non-zero Christoffel Symbols

The Christoffel symbols of the second kind are given by

$$\Gamma^a_{bc} = g^{ad}\left(\partial_c g_{bd} + \partial_b g_{cd} - \partial_d g_{bc}\right)/2, \tag{A.1}$$

where ∂_c denotes $\partial/\partial x^c$ and when applied to the metric given by Eq. (2.4) the non-zero, independent Christoffel symbols become

$$\begin{aligned}
\Gamma^0_{00} &= (a_t c - a_r b + 2bb_t)/2\omega, & \Gamma^0_{01} &= (a_r c - bc_t)/2\omega,\\
\Gamma^0_{11} &= (cc_t - bc_r + 2b_r c)/2\omega, & \Gamma^0_{22} &= br/\omega,\\
\Gamma^0_{33} &= (br\sin^2\theta)/\omega, & \Gamma^1_{00} &= \frac{a_t b + aa_r - 2ab_t}{2\omega},\\
\Gamma^1_{01} &= (ac_t + a_r c)/2\omega, & \Gamma^1_{11} &= \frac{bc_t + ac_r + 2bb_r}{2\omega},\\
\Gamma^1_{22} &= -ar/\omega, & \Gamma^1_{33} &= -(ar\sin^2\theta)/\omega,\\
\Gamma^2_{12} &= 1/r, & \Gamma^2_{33} &= -\cos\theta\sin\theta,\\
\Gamma^3_{13} &= 1/r, & \Gamma^3_{23} &= \cot\theta.
\end{aligned}$$

A.2 Non-zero Mixed Einstein Tensor

The components of the mixed Einstein tensor are obtained using

$$G^a_b = R^a_b - \frac{1}{2}\delta^a_b R, \tag{A.2}$$

J. O'Leary, *General Relativistic and Post-Newtonian Dynamics for Near-Earth Objects and Solar System Bodies*, Springer Theses,
https://doi.org/10.1007/978-3-030-80185-4

where δ^a_b, R^a_b and R denote the Kronecker delta, the Ricci tensor and the Ricci scalar respectively. On application to Eq. (2.4), the non-zero, independent components become

$$G^0_0 = \left[(a^2c_r - a_r b^2 + 2abb_r)r + \omega^2 - a\omega\right]/(\omega r)^2, \tag{A.3}$$

$$G^1_0 = -(a^2c_t + 2abb_t - a_t b^2)/\omega^2 r, \tag{A.4}$$

$$G^0_1 = \left[\omega c_t + (ac_r + a_r c)b + 2b^2 b_r\right]/\omega^2 r, \tag{A.5}$$

$$G^1_1 = -\left\{\left[abc_t + (a_t b + aa_r - 2ab_t)c + a_r b^2\right]r - \omega^2 + a\omega\right\}/(\omega r)^2, \tag{A.6}$$

$$\begin{aligned} G^2_2 = \{&[2\omega c_{tt} - ac_t^2 - (a_t c + a_r b + 2bb_t)c_t + \\ &(aa_r + a_t b - 2ab_t)c_r + (a_r^2 - 2aa_{rr} + 4ab_{rt} - 2a_t b_r)c - \\ &2b(a_r b_r - a_{rr} b + 2b_r b_t + 2bb_{rt})]r - 2a(ac_r + bc_t) + \\ &2(2ab_t - a_t b - aa_r)c + 4b(ab_r - a_r b)\}/4\omega^2 r. \end{aligned} \tag{A.7}$$

A.3 Non-zero Covariant Einstein Tensor

The components of the covariant Einstein tensor are determined from

$$G_{ab} = R_{ab} - \frac{1}{2}g_{ab}R, \tag{A.8}$$

and when applied to Eq. (2.4), the non-zero, independent components are given by

$$\begin{aligned} G_{00} = &\left[a(a^2c_r - a_r b^2 + 2abb_r)r + \omega^2 - a\omega\right]/(\omega r)^2 - \\ &b(a^2c_t + 2abb_t - a_t b^2)/\omega^2 r, \end{aligned} \tag{A.9}$$

$$\begin{aligned} G_{01} = &a(\omega c_t + abc_r + a_r bc + 2b^2 b_r)/\omega^2 r - \\ &\left\{b\left[abc_t + (aa_r + ba_t - 2ab_t)c + b^2 a_r\right]r - \omega^2 + a\omega\right\}/(\omega r)^2, \end{aligned} \tag{A.10}$$

$$\begin{aligned} G_{11} = &c\{\left[abc_t + (aa_r + ba_t - 2ab_t)c + a_r b^2\right]r - \omega^2 + \\ &a\omega\}/(\omega r)^2 + b(\omega c_t + abc_r + bca_r + 2b^2 ba_r)/\omega^2 r, \end{aligned} \tag{A.11}$$

$$\begin{aligned} G_{22} = -r\{&[2\omega c_{tt} - ac_t^2 - (a_t c + a_r b + 2bb_t)c_t + \\ &(aa_r + a_t b - 2ab_t)c_r + (a_r^2 - 2aa_{rr} + 4ab_{rt} - 2a_t b_r)c - \\ &2b(a_r b_r - a_{rr} b + 2b_r b_t + 2bb_{rt})]r - 2a(ac_r + bc_t) + \\ &2(2ab_t - a_t b - aa_r)c + 4b(ab_r - a_r b)\}/4\omega^2. \end{aligned} \tag{A.12}$$

A.4 Integration Formulae

To explicitly determine the expression for ξ from Eq. (2.34), let us consider the expanded expression given by

$$\xi = \tau + \int \frac{r}{r + r_0} dr + \int \frac{[r^2 - \beta(r + r_0)]^{1/2}}{r + r_0} dr, \tag{A.13}$$

where attention will be restricted to the second integral. We begin by making the substitution $x = (r + r_0)^{-1}$ where $dr = -dx/x^2$ so that the second integral in Eq. (A.13) becomes

$$- \int [(xr_0)^2 - x(2r_0 + \beta) + 1]^{1/2} \frac{dx}{x^2}. \tag{A.14}$$

The integral given by Eq. (A.14) becomes

$$\frac{\sqrt{R}}{x} + (r_0 + \beta/2) \int \frac{dx}{x\sqrt{R}} - r_0^2 \int \frac{dx}{\sqrt{R}}, \tag{A.15}$$

where $R = (r_0 x)^2 - (2r_0 + \beta)x + 1$ and to evaluate we consult Gradshteyn and Ryzhik (2014) (page 81 and 84) from which we may deduce

$$\begin{aligned} \int \frac{dx}{x\sqrt{R}} &= -\ln\left[\frac{2 - (2r_0 + \beta)x + 2\sqrt{R}}{x}\right], \\ \int \frac{dx}{\sqrt{R}} &= \frac{1}{r_0} \ln[2r_0\sqrt{R} + 2r_0^2 x - (2r_0 + \beta)], \end{aligned} \tag{A.16}$$

respectively. Furthermore, using the relation

$$\sqrt{R} = \frac{[r^2 - \beta(r + r_0)]^{1/2}}{r + r_0}, \tag{A.17}$$

with Eq. (A.16) we may evaluate Eq. (A.15) completely given by

$$\begin{aligned} \xi = \tau + r + [r^2 - \beta(r + r_0)]^{1/2} - (r_0 + \beta/2) \ln\left\{[r^2 - \beta(r + r_0)]^{1/2} + r - \beta/2\right\} \\ - r_0 \ln\left\{[r^2 - \beta(r + r_0)]^{1/2} - r - \beta(r + r_0)/2r_0\right\} + \xi_0, \end{aligned} \tag{A.18}$$

and the expression for η may be similarly obtained.

Appendix B

B.1 Derivation of Newtonian Orbit Equation

Due to the spherical symmetry associated with the Newtonian gravitational potential $U(r) \equiv \mu/r$, we can, without loss of generality, restrict motion to the equatorial plane ($\theta = \pi/2$). Hence, the Newtonian Lagrangian $\mathcal{L}_n$ associated with a test particle under the influence of a body of mass M is given in polar coordinates by

$$\mathcal{L}_n = \frac{1}{2}\left(\dot{r}^2 + r^2\dot{\theta}^2\right) - \frac{\mu}{r}. \tag{B.1}$$

The Newtonian constants of motion associated with energy $\mathcal{E}_n$ and angular momentum $\mathcal{J}_n$ arise from Eq. (B.1) due to invariance under time translation and rotation and are given by

$$\mathcal{E}_n = \frac{1}{2}\left(\dot{r}^2 + r^2\dot{\theta}^2\right) + \frac{\mu}{r}, \qquad \mathcal{J}_n = r^2\dot{\theta}, \tag{B.2}$$

respectively. Finally, the equations of motion associated with $\mathcal{L}_n$ are obtained from (4.21) and are given by

$$\ddot{r} - r\dot{\theta}^2 = -\frac{\mu}{r^3}, \qquad r\ddot{\theta} + 2\dot{r}\dot{\theta} = 0, \tag{B.3}$$

where we note Eqs. $(\text{B.2})_b$ and $(\text{B.3})_b$ can be shown to be formally identical following a direct integration of $(\text{B.3})_b$. The Newtonian orbit equation can be obtained in two equivalent ways: exploiting the conservation of angular momentum in either $(\text{B.2})_a$ or $(\text{B.3})_a$. The resulting equation is given by

$$\frac{d^2u}{d\theta^2} + u = \mu\mathcal{J}_n^{-2}. \tag{B.4}$$

J. O'Leary, *General Relativistic and Post-Newtonian Dynamics for Near-Earth Objects and Solar System Bodies*, Springer Theses,
https://doi.org/10.1007/978-3-030-80185-4

The solution of Eq. (B.4) is well known and describes the elliptic orbits associated with Newton and Kepler (Ó Mathúna 2008).

B.2 Derivation of General Relativistic Orbit Equation

The Lagrangian $\mathcal{L}_s$ and Hamiltonian $\mathcal{H}_s$ for a test particle in the Schwarzschild field of a spherically symmetric body of mass M have the same value (Franklin 2010). This is due to the fact that there is no classical gravitational potential which leads to an additional conserved quantity given by

$$\mathcal{L}_s = \mathcal{H}_s = \frac{c^2}{2}, \tag{B.5}$$

where the Lagrangian is given by (d'Inverno 1992; Franklin 2010)

$$\mathcal{L}_s = \frac{1}{2}\left[\left(1 - \frac{2\mu}{c^2 r}\right)c^2\left(\frac{dt}{d\tau}\right)^2 - \left(1 - \frac{2\mu}{c^2 r}\right)^{-1}\left(\frac{dr}{d\tau}\right)^2 - r^2\left(\frac{d\theta}{d\tau}\right)^2\right]. \tag{B.6}$$

We denote proper time by τ and restrict motion to the equatorial plane. Again, the conservation of energy $\mathcal{E}_s$ and angular momentum $\mathcal{J}_s$ arise due to the invariance of $\mathcal{L}_s$ under time translations and spatial rotations and are given by

$$\mathcal{E}_s = \left(1 - \frac{2\mu}{r}\right)c^2\frac{dt}{d\tau}, \qquad \mathcal{J}_s = r^2\frac{d\theta}{d\tau}, \tag{B.7}$$

respectively. The Schwarzschild Binet equation is obtained by substituting $(\text{B.7})_a$ and $(\text{B.7})_b$ in (B.5) and (B.6) and making the appropriate Binet transformation. Alternatively, the Schwarzschild Binet equation can be obtained by exploiting the conservation of energy and angular momentum in the radial geodesic equation of motion. It is important to note that both approaches give the same result. Hence, the nonlinear Schwarzschild Binet equation characterising orbits in spherically symmetric spacetimes is given by

$$\frac{d^2u}{d\theta^2} + u = \mu\mathcal{J}_s^{-2} + 3\mu c^{-2}u^2. \tag{B.8}$$

References

Abbassi AH (2001) General Birkhoff's theorem. arXiv:gr-qc/0103103

Abbott B, Abbott R, Adhikari R, Ajith P, Allen B, Allen G, Amin R, Anderson S, Anderson W, Arain M (2009) LIGO: the laser interferometer gravitational-wave observatory. Rep Prog Phys 72(7):076901

Abbott BP, Abbott R, Abbott T, Abernathy M, Acernese F, Ackley K, Adams C, Adams T, Addesso P, Adhikari R (2016) Observation of gravitational waves from a binary black hole merger. Phys Rev Lett 116(6):061102

Abbott BP, Abbott R, Abbott T, Acernese F, Ackley K, Adams C, Adams T, Addesso P, Adhikari R, Adya V (2017) GW170817: observation of gravitational waves from a binary neutron star inspiral. Phys Rev Lett 119(16):161101

Allan DW, Weiss MA, Ashby N (1985) Around-the-world relativistic Sagnac experiment. Science 228(4695):69–70

Ashby N (2003) Relativity in the global positioning system. Living Rev Relativ 6(1):1

Ashtekar A (2005) 100 years of relativity: space-time structure: Einstein and beyond. World Scientific, Singapore

Barack L, Cardoso V, Nissanke S, Sotiriou TP, Askar A, Belczynski C, Bertone G, Bon E, Blas D, Brito R (2019) Black holes, gravitational waves and fundamental physics: a roadmap. Class Quantum Gravity 36(14):143001

Battista E, Esposito G, Dell'Agnello S (2017) On the foundations of general relativistic celestial mechanics. Int J Mod Phys A 32(26):1730022

Baumgarte TW, Shapiro SL (2010) Numerical relativity: solving Einstein's equations on the computer. Cambridge University Press, Cambridge

Birkhoff G (1923) Relativity and modern physics, vol 1. Harvard University Press, Cambridge

Blanchet L, Damour T (1989) Post-Newtonian generation of gravitational waves. Ann Inst Henri Poincaré A 50:377–408

Blanchet L, Spallicci A, Whiting B (2011) Mass and motion in general relativity, vol 162. Springer, Berlin

Brumberg V (2004) On relativistic equations of motion of an earth satellite. Celest Mech Dyn Astron 88(2):209–225

Brumberg V (2007) On derivation of EIH (Einstein-Infeld-Hoffman) equations of motion from the linearized metric of general relativity theory. Celest Mech Dyn Astron 99(3):245–252

Brumberg V (2017) Essential relativistic celestial mechanics. Routledge, Milton Park

J. O'Leary, *General Relativistic and Post-Newtonian Dynamics for Near-Earth Objects and Solar System Bodies*, Springer Theses,
https://doi.org/10.1007/978-3-030-80185-4

Brumberg V, Groten E (2001) IAU resolutions on reference systems and time scales in practice. Astron Astrophys 367(3):1070–1077

Butcher JC (1964) Implicit Runge-Kutta processes. Math Comput 18(85):50–64

Butcher JC (2016) Numerical methods for ordinary differential equations. Wiley, Hoboken

Chandrasekhar S (1965) The post-Newtonian equations of hydrodynamics in general relativity. Astrophys J 142:1488

Chandrasekhar S (1969) Conservation laws in general relativity and in the post-Newtonian approximations. Astrophys J 158:45

Chandrasekhar S, Nutku Y (1969) The second Post-Newtonian equations of hydrodynamics in general relativity. Astrophys J 158:55

Chandrasekhar S, Thorne KS (1985) The mathematical theory of black holes. Oxford University Press, Oxford

Charon JE (1963) Quinze leçons sur la relativité générale avec une introduction au calcul tensoriel. Kister

Cheng T (2005) Relativity, gravitation and cosmology: a basic introduction. Oxford University Press, Oxford

Chou C, Hume D, Rosenband T, Wineland D (2010) Optical clocks and relativity. Science 329(5999):1630–1633

Ciufolini I, Paolozzi A, Paris C (2012) Overview of the LARES Mission: orbit, error analysis and technological aspects. J Phys: Conf Ser, vol 354. IOP Publishing

Contopoulos G (1990) Periodic orbits and chaos around two black holes. Proc R Soc Lond A 431(1881):183–202

Cooper G (1987) Stability of Runge-Kutta methods for trajectory problems. IMA J Numer Anal 7(1):1–13

Dalarsson M, Dalarsson N (2015) Tensors, relativity, and cosmology. Academic, Cambridge

Damour T (1987) The problem of motion in Newtonian and Einsteinian gravity. In: Hawking S, Israel W (eds) Three hundred years of gravitation, chapter 6. Cambridge University Press, Cambridge, pp 128–198

Damour T, Deruelle N (1985) General relativistic celestial mechanics of binary systems. I. The post-Newtonian motion. Ann Inst Henri Poincaré 43(1):107

Damour T, Soffel M, Chongming, and Xu, (1992) The general relativistic N-body problem. In: Hawking S, Israel W (eds) Relativistic gravity research with emphasis on experiments and observations. Lecture notes in physics. Springer, Berlin, pp 46–69

Damour T, Soffel M, Xu C (1991) General-relativistic celestial mechanics. I. Method and definition of reference systems. Phys Rev D 43(10):3273

Damour T, Soffel M, Xu C (1992) General-relativistic celestial mechanics II. Translational equations of motion. Phys Rev D 45(4):1017

Damour T, Soffel M, Xu C (1993) General-relativistic celestial mechanics. III. Rotational equations of motion. Phys Rev D 47(8):3124

Damour T, Soffel M, Xu C (1994) General-relativistic celestial mechanics. IV. Theory of satellite motion. Phys Rev D 49(2):618

Darwin CG (1959) The gravity field of a particle. Proc R Soc Lond A 249(1257):180–194

De Sitter W (1916) On Einstein's theory of gravitation and it's astronomical consequences. Second paper. Mon Not R Astron Soc 77:155–184

d'Inverno R (1992) Introducing Einstein's relativity. Oxford University Press, USA

Donnelly D, Rogers E (2005) Symplectic integrators: an introduction. Am J Phys 73(10):938–945

Drake S (2006) The equivalence principle as a stepping stone from special to general relativity: a socratic dialog. Am J Phys 74(1):22–25

Drake S, Dettmann C, Frankel N, Cornish N (1996) Chaos in special relativistic dynamics. Phys Rev E 53(2):1351

Droste J (1917) The field of a single centre in Einstein's theory of gravitation, and the motion of a particle in that field. Ned Acad Wet, SA 19:197

Eddington AS (1920) The mathematical theory of relativity. Cambridge University Press, Cambridge

Eddington AS (1924) A Comparison of whitehead's and Einstein's formulae. Nature 113(2832):192

Eddington AS, Clark GL (1938) The problem of n bodies in general relativity theory. Proc R Soc Lond A 166(927):465–475

Einstein A (1916a) Approximative integration of the field equations of gravitation. Sitzungsber. Preuss. Akad. Wiss. Berlin (Math Phys) 1916:688–696

Einstein A (1916b) Näherungsweise Integration der Feldgleichungen der Gravitation. Sitzungsberichte der Königlich Preußischen Akademie der Wissenschaften (Berlin) 688–696

Einstein A (1918) Über gravitationswellen. Sitzungsberichte der Königlich Preußischen Akademie der Wissenschaften (Berlin). Seite, p. 154–167

Einstein A, Infeld L, Hoffmann B (1938) The gravitational equations and the problem of motion. Ann Math 65–100

Event Horizon Telescope Collaboration (2019). First M87 event horizon telescope results. I. The shadow of the supermassive black hole. arXiv:1906.11238

Finkelstein D (1958) Past-future asymmetry of the gravitational field of a point particle. Phys Rev 110(4):965

Florides PS, Synge JL (1961) Notes on the schwarzschild line-element. Report, Dublin Inst. for Advanced Studies

Fock V (1964) The theory of space, time and gravitation. MacMillan, New York

Folkner WM, Williams JG, Boggs DH, Park RS, Kuchynka P (2014) The planetary and lunar ephemerides DE430 and DE431. Interplanetary Network Progress Report 196:1–81

Foster J, Nightingale JD (2010) A short course in general relativity. Springer Science & Business Media

Franklin J (2010) Advanced mechanics and general relativity. Cambridge University Press, Cambridge

Goldstein A, Veres P, Burns E, Briggs M, Hamburg R, Kocevski D, Wilson-Hodge C, Preece R, Poolakkil S, Roberts O (2017) An ordinary short gamma-ray burst with extraordinary implications: fermi-GBM detection of GRB 170817A. Astrophys J Lett 848(2):L14

Goldstein H (1965) Classical mechanics. Pearson Education India

Gradshteyn IS, Ryzhik IM (2014) Table of integrals, series, and products. Academic, Cambridge

Griffiths JB, Podolskỳ J (2009) Exact space-times in Einstein's general relativity. Cambridge University Press, Cambridge

Hafele JC, Keating RE (1972) Around-the-world atomic clocks: predicted relativistic time gains. Science 177(4044):166–168

Hagihara Y (1930) Theory of the relativistic trajectories in a gravitational field of Schwarzschild. Jpn J Astron Geophys 8:67

Hairer E, Lubich C, Wanner G (2006) Geometric numerical integration: structure-preserving algorithms for ordinary differential equations, vol 31. Springer Science & Business Media

Harry GM, Collaboration LS (2010) Advanced LIGO: the next generation of gravitational wave detectors. Class Quantum Gravity 27(8):084006

Hartle JB, Hawking SW (1972) Solutions of the Einstein-Maxwell equations with many black holes. Commun Math Phys 26(2):87–101

Hawking SW, Ellis GFR (1973) The large scale structure of space-time, vol 1. Cambridge University Press, Cambridge

Hawking SW, Israel W (1989) Three hundred years of gravitation. Cambridge University Press, Cambridge

Hill JM, O'Leary J (2018) Generalized transformations and coordinates for static spherically symmetric general relativity. R Soc Open Sci 5(4):171109

Hobson MP, Efstathiou GP, Lasenby AN (2006) General relativity: an introduction for physicists. Cambridge University Press, Cambridge

Hugentobler U (2010) Supporting material: Orbit perturbations due to relativistic corrections. In: Petit G, Luzum B (eds) IERS conventions (2010), Chapter 10. IERS Technical Notes

Johansen NV, Ravndal F (2006) On the discovery of Birkhoff's theorem. Gen Relativ Gravit 38(3):537–540

Kerr RP (1963) Gravitational field of a spinning mass as an example of algebraically special metrics. Phys Rev Lett 11(5):237–238

Kobayashi M, Yamada K (2019) Spacecraft orbit around two fixed bodies. Acta Astronautica

Koberlein B, Meisel D (2013) Astrophysics through computation: with mathematica®. Cambridge University Press, Cambridge

Kopeikin S, Efroimsky M, Kaplan G (2011) Relativistic celestial mechanics of the solar system. Wiley, Hoboken

Kopeikin SM (2018) Carter-like Constant of motion in newtonian gravity is the vinti integral. arXiv:1806.04547

Landau LD (2013) The classical theory of fields, vol 2. Elsevier, Amsterdam

Landau LD, Lifshitz EM (2013) Course of theoretical physics: mechanics, vol 1. Elsevier, Amsterdam

Lichnerowicz A, Teichmann T (1955) Théories relativistes de la gravitation et de l'électromagnétisme. Phys Today 8:24

Lorentz HA, Droste J (1937). The motion of a system of bodies under the influence of their mutual attraction, according to Einstein's Theory. In: Collected papers: volume V. Springer, Dordrecht, pp 330–355

Majumdar SD (1947) A class of exact solutions of Einstein's field equations. Phys Rev 72(5):390

Marty J (2013) Algorithmic documentation of the GINS software. GINS algorithm overview

Maxima (2016) Maxima, a computer algebra system version 5.38.1

McLachlan Robert I, Quispel GRW (2006) Geometric integrators for ODEs. J Phys A: Math Gen 39(19):5251

Meegan C, Lichti G, Bhat P, Bissaldi E, Briggs MS, Connaughton V, Diehl R, Fishman G, Greiner J, Hoover AS (2009) The Fermi gamma-ray burst monitor. Astrophys J 702(1):791

Mirshekari S, Will CM (2010) Carter-like constants of motion in the Newtonian and relativistic two-center problems. Class Quantum Gravity 27(23):235021

Misner CW, Thorne KS, Wheeler JA (1973) Gravitation. Macmillan, New York

Montenbruck O, Gill E (2012) Satellite orbits: models, methods and applications. Springer Science & Business Media

Moyer TD (1971) Mathematical formulation of the double precision orbit determination program/DPODP. Technical report

Moyer TD (2005) Formulation for observed and computed values of deep space network data types for navigation, vol 3. Wiley, Hoboken

Müller J, Soffel M, Klioner SA (2008) Geodesy and relativity. J Geodesy 82(3):133–145

Müller T, Grave F (2009) Catalogue of spacetimes. arXiv:0904.4184

Musielak ZE, Quarles B (2014) The three-body problem. Rep Prog Phys 77(6):065901

Nordtvedt K (1968) Equivalence principle for massive bodies. II. Theory. Phys Rev 169(5):1017–1025

Ó Mathúna D (2008) Integrable systems in celestial mechanics, vol 51. Springer Science & Business Media

Ohanian HC, Ruffini R (2013) Gravitation and spacetime. Cambridge University Press, Cambridge

O'Keefe JA (1957) An application of Jacobi's integral to the motion of an earth satellite. Astron J 62:265

O'Leary J, Hill JM (2018) Post-newtonian satellite orbits. Astrophys Space Sci 363(10):211

O'Leary J, Hill JM (2018) Numerical conservation of exact and approximate first post-newtonian energy integrals

O'Leary J, Hill JM, Bennett JC (2018) On the energy integral for first post-newtonian approximation. Celest Mech Dyn Astron 130(7):44

Pail R (2014) CHAMP, GRACE, GOCE-Satellite Projects. In: Grafarend E (ed) Encyclopedia of geodesy. Springer International Publishing, Cham, pp 1–11

Papapetrou A (1947) A static solution of the equations of the gravitational field for an arbitrary charge-distribution. In: Proceedings of the royal irish academy. Section A: mathematical and physical sciences, vol 51, pp 191–204

Peron R, Colpi M, Gorini V, Moschella U (2015) Gravity: where do we stand? Springer

Petit G, Luzum B (2010) IERS conventions (2010). Report, DTIC Document

Pireaux S, Barriot J-P, Rosenblatt P (2006) SCRMI: A (S)emi-(C)lassical (R)elativistic (M)otion (I)ntegrator to model the orbits of space probes around the Earth and other planets. Acta Astronaut 59(7):517–523

Pireaux S, Chauvineau B, Hees A (2008) Relativistic versus Newtonian orbit model: the Relativistic Motion Integrator (RMI) software. Illustration with the LISA mission. arXiv:0801.3637

Poisson E (2007) Post-Newtonian theory for the common reader. Lecture Notes

Poisson E, Will CM (2014) Gravity: Newtonian, post-Newtonian, relativistic. Cambridge University Press, Cambridge

Puetzfeld D, Lämmerzahl C (2019) Relativistic geodesy. Fundamental theories of physics

Richardson D, Kelly T (1988) Two-body motion in the post-Newtonian approximation. Springer, Berlin

Rodríguez CM (1987) Orbits in general relativity: the Jacobian elliptic functions. Il Nuovo Cimento B (1971-1996) 98(1), 87–96

Roh K, Choi B (2014) The effects of the IERS conventions (2010) on high precision orbit propagation. J Astron Space Sci 31(1):41–50

Roh K, Kopeikin SM, Cho J (2016) Numerical simulation of the post-Newtonian equations of motion for the near Earth satellite with an application to the LARES satellite. Adv Space Res 58(11):2255–2268

Schutz B (2009) A first course in general relativity. Cambridge University Press, Cambridge

. Schwarzschild K (1916) On the gravitational field of a point mass according to Einstein's theory. Sitzungsber. Preuss. Akad. Wiss., Phys Math

Soffel M (1989) Relativity in astrometry, celestial mechanics and geodesy. Springer, Berlin

Soffel M (2014) On the DSX-framework. In: Kopeikin SM (ed) Frontiers in relativistic celestial mechanics: Theory, Volume 1, Chapter 4. De Gruyter

Soffel M, Klioner SA, Petit G, Wolf P, Kopeikin S, Bretagnon P, Brumberg V, Capitaine N, Damour T, Fukushima T (2003) The IAU 2000 resolutions for astrometry, celestial mechanics, and metrology in the relativistic framework: explanatory supplement. Astron J 126(6):2687

Soffel M, Langhans R (2012) Space-time reference systems. Springer Science & Business Media

Soffel MH, Han W-B (2019) Applied general relativity. Astronomy and Astrophysics Library, Springer Nature Switzerland, Cham

Stephani H, Kramer D, MacCallum M, Hoenselaers C, Herlt E (2009) Exact solutions of Einstein's field equations. Cambridge University Press, Cambridge

Straumann N (2004) General relativity with applications to astrophysics

Suzuki H, Gupta P, Okawa H, Maeda K-I (2019) Cumulative shift of periastron time of binary pulsar with Kozai-Lidov oscillation. Mon Not R Astron Soc: Lett 486(1):L52–L57

Thorne KS, Will CM (1971) Theoretical frameworks for testing relativistic gravity. I. Foundations. Astrophys J 163:595

Tucker A, Will CM (2019) Pericenter advance in general relativity: comparison of approaches at high post-Newtonian orders. Class Quantum Gravity 36(11):115001

Vallado DA (2001) Fundamentals of astrodynamics and applications, vol 12. Springer Science & Business Media

Vinti JP (1959) A new method of solution for unretarded satellite orbits. National Bureau of Standards, United States Department of Commerce

Viswanathan V, Fienga A, Minazzoli O, Bernus L, Laskar J, Gastineau M (2018) The new lunar ephemeris INPOP17a and its application to fundamental physics. Mon Not R Astron Soc 476(2):1877–1888

Wald RM (1984) General relativity. University of Chicago Press, Chicago

Weinberg S (1972) Gravitation and cosmology: principles and applications of the general theory of relativity, vol 67. Wiley, New York

Whittaker ET (1988) A treatise on the analytical dynamics of particles and rigid bodies. Cambridge University Press, Cambridge

Will CM (1971a) Theoretical frameworks for testing relativistic gravity. II. Parametrized post-Newtonian hydrodynamics, and the Nordtvedt effect. Astrophys J 163:611

Will CM (1971b) Theoretical frameworks for testing relativistic gravity. III. Conservation laws, Lorentz invariance, and values of the PPN parameters. Astrophys J 169:125

Will CM (2009) Carter-like constants of the motion in Newtonian gravity and electrodynamics. Phys Rev Lett 102(6):061101

Will CM (2011) On the unreasonable effectiveness of the post-Newtonian approximation in gravitational physics. Proc Natl Acad Sci 108(15):5938–5945

Will CM (2014) Incorporating post-Newtonian effects in N-body dynamics. Phys Rev D 89(4):044043

Will CM (2018) New general relativistic contribution to Mercury's perihelion advance. Phys Rev Lett 120(19):191101

Will CM, Nordtvedt K (1972) Conservation Laws and Preferred Frames in Relativistic Gravity. I. Preferred-Frame Theories and an extended PPN Formalism. Astrophys J 177:757

Wintner A (1941) The analytical foundations of celestial mechanics. Princeton University Press, Princeton; Oxford University Press, H. Milford, p 1941